IGOR KONOVALOV

Reciprocalism

What does sustainability actually mean?

First published by Igor Konovalov (independent) 2025

First edition

Cover art by Diren Yardimli

This book was professionally typeset on Reedsy.
Find out more at reedsy.com

To my fiancée Alla;
my intellectual sparring partner,
my support,
my energy source.

Contents

Introduction

The book that you, my dear reader, are currently holding provides an in-depth analysis of the word 'sustainability'. This research stems from my passion for the environmental sciences, but also from my frustration of the use and misuse, often even abuse, of the term. Businesses, for instance, boldly proclaim their switch to single-ply toilet paper to save the rainforest, while their core activities produce private jets and abuse employees – not very sustainable, I would say. Another example are governments that minimise funding for renewable energy initiatives while subsidising fossil fuels and dare to sell their policies to the public as matters of national security or autonomy.

You can probably sense my frustration from these examples; I have my own personal view on these matters. But none of these actors themselves claim to be evil, even though I might often perceive some of them as such. To the contrary, they all claim that they are doing the correct thing – the single correct thing. They are all right in their own perception. But am I?

I find this a fascinating question: Why do I think of all matters involving sustainability and the environment in such a different way than many others? There are three possible answers. First, either I am right and most others are wrong. In which case, why? Second, I am wrong and most others are right. In which case, how could I have been led so astray? Third – the most likely answer – we are all partially right. The truth has a tendency to lie in the middle, as they say. If this is the case, are we then not fighting each other while looking at different sides of the same coin?

It is worth mentioning that my research was not originally focussed on these questions. I had spent years attempting to understand the concept of biodiversity. This was an endeavour that started as a hobby but quickly grew out of control. It started several years ago when I, still very much a junior consultant at the time, asked a senior colleague why biodiversity was not a part of the global standardised method for the calculation of environmental impacts. 'Probably too difficult', was the answer. Being young and naïve, I found this a weak excuse. How difficult could it be? Well, it turns out, very. It took four years of failed attempts to determine a method of calculation that could include biodiversity, before realising that any result depended heavily on various assumptions, for instance, whether we should consider the environmental impacts on the oceans to be worse than those on forests. So as my confusion and frustration grew, there came a point where I threw down my digital pen and metaphorically burned my spreadsheets. It seemed that quantifying biodiversity was impossible! But as I said, this endeavour of mine got out of control. Although I had absolutely nothing to show for the hours I had already spent, I could not let it go that easily.

So I started again. This time, however, I did not try to determine the impact of human activities on biodiversity, but rather to answer the question as to what actually constitutes impact on biodiversity. This turned out to be a highly complicated question, even though it can appear simple at first glance. In terms of human activities which impact biodiversity, planting a tree would be good, throwing a plastic bag in the river would be bad. Simple, right? But if we were to investigate, for example, The Ocean Cleanup and their daunting task to rid the oceans of plastic pollution[1] – in which they are using large, fossil-fuel-powered ships to collect that plastic waste – and scooping up bundled plastic in which sea creatures might be hiding, we may wonder whether the side effects of the medicine are a worthwhile payoff to the cure of the main disease. To be clear, I highly respect The Ocean Cleanup and their founder. But it does illustrate the fact that human impact is always present. It does not hurt to study the potential side effects of human activities, no matter how well-intended these activities are.

As already stated, the question of what actually constitutes impact on biodiversity is complicated. And yet, you are about to read a book which shows that this research has actually been completed. I promise you, it actually leads to a fascinating conclusion, one which would not have been achievable without the groundwork conducted by Graham Harman and Timothy Morton on the subject of Object-Oriented Ontology (OOO). Although these two scholars are not unique in working on OOO, they are the ones that have had the most impact on my work. It is through their works, most notably Morton's *All art is ecological* (2021) and Harman's *Object-Oriented Ontology, A New Theory*

---

[1] Started as a thesis project of Boyan Slat, The Ocean Cleanup is an organisation that recognised that plastic pollution in oceans over time clustered itself in several hotspots, so called gyres, due to ocean currents. Instead of attempting to remove plastic pollution throughout the entire ocean, it is much more effective to target cleaning efforts toward these gyres. For more than a decade, The Ocean Cleanup has developed and implemented systems to collect and remove plastic from ocean gyres and rivers.

of Everything (2018), that I got acquainted with OOO. And it is through OOO that I found the answers to my questions.

Before I elaborate on the principles of OOO, you might be wondering why I am talking about biodiversity while the subtitle lured you in with the promise of a discussion on the subject of sustainability. But they are very much related to each other. Think of it like this: When discussing sustainability, most likely the burning of fossil fuels and global warming come to mind first. We should stop burning fossil fuels because they are heating up the planet. But is the negative impact actually the heating of the planet? Well, not exactly. It is the effect that this heating causes on all forms of life, better yet all objects, on the planet. An example of such effect could be the increase in droughts and floods that decimate ecosystems and all creatures that live within them. Sustainability thus directly relates to biodiversity. Similarly if we look at the example of ocean pollution. It is not the oceans containing plastic waste that is the problem, but rather the many creatures that accidentally ingest, get entangled or suffocate from it, to be precise.

This is how I ended up drifting away from a focus on biodiversity, and ended up realising that I was actually writing on the basics of sustainability. This is the point where OOO started to become fundamental to my research. One of the foundational ideas of OOO is the consideration that reality extends beyond the human mind. Our understanding, or lack thereof, of something does not dictate whether or not that something exists. Due to this standpoint, OOO does not assert a hierarchy upon objects. It does not place humans above anything else, which is typical for many other philosophical movements. Instead, it starts from a flat baseline. Everything is equally important. This is why philosophers within OOO tend to speak in terms of 'objects'. A human is an object just as much as a mouse, a tree, the wind or anything else; be it material or non-material.

This baseline turns out to be ideal when dealing with sustainability. Since a seal getting entangled in plastic pollution cannot be considered worse than a field scorching under a heatwave caused by the global temperature increasing by 1.5 degree Celsius above pre-industrial level, one can actually investigate them for what they are: impacts of humans on the objects around them.[2] However, this still left me with one question that was gnawing at my confidence. Will being alive not always lead to some form of impact on one's surroundings? Even planting a tree, the perceived pinnacle of sustainable behaviour, means that you have to spend resources such as water and that you will change the landscape. If every instant of a human life will inevitably result in some kind of impact on other objects, is there actually such a thing as sustainability?

To give you a very brief summary of the entirety of the book: yes! There is such a thing as sustainability. And although I do not want to spoil the ending, I will explain to you how I came to this conclusion, as this also explains how the book is structured. What you will find here is a series of chapters which can be read as a collection of essays. Each essay covers its own topic while building on the insights gathered in the preceding essays. At the end of most of these you will find a very brief descriptor of the most important insight from that section. Put

---

[2] The average global temperature in 2024 exceeded the 1.5 degrees Celsius increase since pre-industrial levels, considered to be the safe limit by the International Panel on Climate Change. Although this was a measurement of a single year and can thus not be directly related to the global climate (which is the temperature averaged over decades), it does indicate an alarming development.

together, I call these the 'fundamental rules of sustainability'. These rules, often derived from the principles of OOO, describe something which I deem universal among all objects. This set of rules, which you can find summarised at the very end of the book, starts for instance with every object having wants and needs, for which the object needs other objects in order to fulfil them. The fundamental rules of sustainability together define the framework from which a single definition for the concept of sustainability is constructed.

Now, if you are like me and consider yourself a person with a practical mindset, you might be wondering if I ended my research by defining the term sustainability, or if there are further implications. You are in luck: there is! And it is called 'reciprocalism'. Much like socialism, impressionism, feminism and all the other 'isms', reciprocalism indicates a specific mindset around a common belief. In the case of reciprocalism, that belief is the strive for as much sustainability as possible. It is my suggestion for a new 'ism', which should enable you to pursue sustainability within the context of your own life.

Before I send you on the path that should guide you through the forest that is my research, I wish to explain to you one more detail. As already mentioned, OOO speaks of objects. I did however give my own spin to the usage of the term 'objects'. And although an explanation of my usage of the term can be found in the first chapter, I do wish to give you a brief summary of the three types of objects I classified. These I call individual objects, collective objects and coupled objects. Individual objects are a single entity, such as one person. Collective objects are a collection of individual objects, such as a football team consisting of several individual players. You can see however that a collective object can also turn into an individual object depending on how it is described. A football team becomes an individual object when we are interested in the team as a whole, rather than the separate members. Individual and collective objects are hence interchangeable and depend on the context of the analysis. The coupled object is the link between two or more objects. For instance, the player as part of the football team joins the team because they might feel a sense of social obligation. Regardless, we are no longer describing the player as an individual, nor are we talking about the team. Instead, we are talking about the link between the player and the team. These three types of objects are extensively described and utilised in the book. And all of them can take any form; living or non-living, physical entities, actions, thoughts, and much more.

And with that, there is nothing left for me to do but wish that you enjoy reading *Reciprocalism*.

1

A perspective of one and many

Sufficiency of autonomy, duty of community

If there is one sentence that has stuck with me my entire life, it must be what my mother told me when I was young: 'Too' is never good. Whether it refers to too little or too much, too hot or too cold, too rich or too poor. One should strive for balance. The underlying philosophy was that you should be content if you could live a burden-free life. And this balance implied a range wherein you would live your best life: Not too many or too crazy parties, but still enjoying time with friends. Not every day alone at home, but enjoying time on your own every now and again.

This range is what I call 'personal sufficiency'. As an individual I have certain needs, such as food and water which I simply could not live without. And I have certain wants, such as herring and sparkling lemonade which make life enjoyable. But eat herring every day and your kidneys will struggle with the salt; drink lemonade every day and your teeth will dissolve in the acid from the carbonated water. I understand from this that one's needs ensure survival whereas one's wants ensure that they live.

Something I occasionally want, and one might argue that I might need them for a healthy diet, are mushrooms. But what of the wants and needs to the mushroom itself? *Boletus edulis*, a species of edible mushroom frequently spotted in parks and forests across the Netherlands, needs a climate in the range from subtropical to temperate and a forest ecosystem to survive. It appears to prefer conifer trees and a slightly cooler temperature above other tree species and the upper survivable temperature limit. At least that is where more fruiting bodies seem to be located.[3]

For the sake of discussion, we can take a detour to the realm of the bearded vulture, *Gypaetus barbatus*, and investigate some of its wants and needs. It mostly, but not exclusively, eats the bones from deceased mammals, ungulates to be precise. That appears to be its want. But it needs dead animals in general, be it mammals, birds or reptiles.

This distinction between want and need appears to be able to be made for all living organisms. But it can also be extended to non-living entities. A non-living object can for instance be a river. A river needs elevation and a sufficient supply of water, which will enable the required movement of water. Without

---

[3] An interesting article by Hall et al. called '*Ectomycorrhizal fungi with edible fruiting bodies - 2. Boletus edulis*' (1998) found that fruiting of *Boletus edulis* depended primarily on temperature and soil moisture and occurred mostly when temperatures started to decline after a long and hot period (i.e. autumn). Although cosmopolitan in distribution, this mushroom species tends to develop fruiting bodies in September in the UK and in November in California.

elevation a river would simply be a lake, and without enough water a river would be a dry rock cascade. But a river also wants a distinct lack of hurdles. Place a barrier in a river and it will find a way around it, as if it needs to keep flowing.

Consider a rock tumbling in the river. It needs geothermal heat to form from its constituent elements deep inside the Earth's layers and water to erode it from the mountain from which it rose to the surface. It wants sufficient water to break free from the mountain, but not too much as to erode it into oblivion.

These examples have very briefly touched upon the nature of the individual object. Matters change however when examining a collective object. Whereas the rock needs water to break free from the mountain, the mountain does not want water because it threatens its structural integrity. The vulture wants bone marrow from ungulates, but a group of vultures may need to make do with meat from rodents. A mushroom wants conifer trees, but a group of mushrooms needs many different woody plants.

Similarly, I want to spend my time cooking for my partner but society needs me working a full work week. A collective object can thus have different wants and needs to the individual object. It is a different object after all. And by allowing each object its needs and the occasional wants, it appears to be easy to provide each with its best life. Matters start getting complicated when the individual object and the collective object are intertwined, as for instance a person and the society in which they live. A person in a society is expected to work full-time to provide services to the community, while wanting to occasionally go to a movie theatre to relax. In order for our person to relax however, the society needs to provide the movie theatre, for which our person needs to provide resources to the community in turn. This is the difference between the wants and needs of an individual and a collective object. Individual objects need to provide to the collective object of which they are a part, in order to get what they want. A mushroom wants to be a certain distance from other mushrooms as it helps with spore distribution. But it needs to be close enough if it wants to remain part of the mycelium network and gain the beneficial nutrients from it. The same holds true for the vulture. It wants to be sufficiently far away from other vultures so as to not be required to share its meals. But it needs to remain connected to the larger collective of vultures when searching for a mate. Even the river, which wants to flow freely from higher to lower elevation, needs to remain connected to the upstream aquifer system if it wants to keep receiving the water essential to its survival.

What I take from this brief discussion is that individual objects – specifically objects consisting as part of a collective, defined as coupled objects – can have their wants and needs fulfilled only by supplying the collective with what it needs. But we can also reason the inverse of this conclusion. Collective objects can only have their wants and needs met by providing for the wants and needs of individual objects. A collective wants me to work full-time for which it needs to provide me with an environment free of conflict. A mycelium wants all mushrooms to be tied together through the underground network, for which it needs to supply every mushroom in the entire network with nutrients. A collective of vultures wants breeding pairs, for which it needs to allow individuals to exist within relatively close proximity without fighting over food.[4] An aquifer system wants to remain intact and not be affected by erosion, so it needs to allow the river to flow with its excess water. And the mountain wants to grow to the sky as a result of tectonic activity, but for this it needs to shed the occasional rock to prevent itself from collapsing.

Arguably, some of these examples might seem conceptually far-fetched. But this is only logical as it attempts to escape the framework of OOO wherein every object has its own reality which cannot be fully grasped by another object.[5] I myself am an object. Hence, I can understand the mountain only so far as my own reasoning carries me. Nonetheless, it does show me that all objects – be they individual, collective or coupled – supplement each other in a certain manner when it comes to providing all participants' wants and needs.

Let us explore the coupled object more in depth. This is an object that is very tightly intertwined with another object. Another way of looking at this would be to consider the link itself as the object. This can be an individual object interlinked with a collective object, such as the exchange of labour for a stable living environment free of conflict. But it can also be a collective object interlinked with an individual object, such as a field of grass consisting of many individual grass organisms,

[4] One could argue that a collective of vultures needs breeding pairs in order to ensure the survival of the collective. I have instead chosen to describe it as a want, since the survival of the collective does not require every vulture to form a breeding pair. As long as there are enough chicks born, the collective will survive. Does this deem breeding pairs a want or a need to the collective? Perhaps both.

[5] How we can understand other objects' wants and needs is discussed extensively throughout this book, most notably in chapter 5.

with a cow. In the former example, the coupled object would be. The latter example can be considered as an exchange of food for the cow and a clearing of trees and shrubs for the grass.

Aside from the coupled object linking the intertwined individual and collective objects, it can also function as a link between individual objects, or between collective objects. When I go to the fish shop, the owner gets my money and I receive a fish. And when a group of Brazilian dolphins drive fish to a group of Brazilian fishermen, both gain fish without spending the energy they would have spent when operating as a decoupled object. In this latter example, the fishermen have to accept a part of the catch going to the dolphins.

The coupled object can thus be understood as a collaboration. An object provides to the wants and needs of another object, and this second object provides to the wants and needs of the first object. In addition a coupled object can also provide to the wants and needs of another coupled object. Consider, for example, the Yellowstone National Park in the United States, specifically the relation between the wolf and raven on the one hand, and the beaver and willow on the other. The raven leads wolves to a location with deer, their potential prey, which when killed by the wolves form a source of food for both the wolves and the raven. The wolves had to spend less effort finding the prey, while the raven got access to food that they would otherwise not have been able to obtain. Both have thus been providing to each other's wants and needs. But in doing so they remove the grazing pressure from these deer on willow plants. Without the wolves aided by the raven, these willows were being badly damaged by deer, as has been documented in the few decades between the killing of the last Yellowstone wolf and their reintroduction.[6] These willows in turn provide a source of food and material for beavers. And returning the favour, the beavers thin out the willows without trampling the willow collective, as the deer would do. Hence, the wolf-raven interaction benefits in a similar manner to the willow-beaver interaction.

These examples bring to mind the old adage of the whole being greater than the sum of its parts. And it makes up a large part of the environmental movement's philosophy that all of the Earth's inhabitants are linked and dependent upon each other in some way. From this we can derive the first rule of sustainability.

(1)

An object, of whatever form, is inherently linked to other objects, and depends on interacting with other objects to provide its own wants and needs.

This first rule of sustainability is important. Interaction between objects is imperative for their existence. Even if you would place a human in a hypothetical place fully devoid of other objects, the mere existence of the human demands other

---

[6] Several studies were found that arrived at similar conclusions that the absence of wolves in Yellowstone caused a larger population of deer, which in turn resulted in more willow being trampled, which in turn heavily impacted the population of beavers. My favourite article, *Hydrologic regime and herbivory stabilize an alternative state in Yellowstone National Park (2007)* - very aptly authored by Evan Wolf et al. - went a step further and suggested that simply reducing the grazing pressure of deer alone may not yield the historical state of the landscape since waterways will have changed. It implies that landscapes are likely to be permanently altered by human impacts, even if those impacts are eventually reversed.

objects (e.g. breathable air) without which he would perish, which follows the rule as outlined above.

A society of societies

Taking the first rule of sustainability discussed above, one might be interested to understand what practical meaning this has to human civilisation. After all, we want to derive an understanding of sustainability in order to restore the massive impact humanity poses on the planet to a more natural level. And since I am writing this document for a human audience, rather than one of mushrooms or vultures, I wish to spend some effort to understand the implications of the necessity of interactions between objects on modern human life.

Let us break down society to its constituent components and start our analysis with a single person. This person has needs of food, water, clothing and shelter. This person also has many wants, but we will investigate them later. Taking account of this person's needs, we can imagine a situation where all the needs can be fulfilled by the individual person himself. They can be skilled enough at farming to provide enough food, knowledgeable on the matter of water treatment to provide drinking water, and capable of making clothes, building a house and creating all the necessary tools for these tasks. It might be very hard work but theoretically this person could provide in all these needs by themselves.

But what if a severe storm passes over this person's farm, destroying a portion of their crops? The risk that our person would starve has suddenly increased significantly. And what about a lightning strike which starts a fire destroying their house? Or an insect infestation destroying their clothes? Or any of a plethora of risks that could diminish a water supply? All are external factors that will impact our person's chances of survival.

Now imagine the same person living within a small community. Again the house is struck by lightning and burns down. Instead of having to reside in the harsh outdoors, our person can now take temporary shelter with one of the other residents of the village until their house has been rebuilt. And if the harvest is destroyed by a storm, it will be quicker to replant when collaborating with others.

The case of water shortages due to drought will however still pose a significant peril to our person within their village. As drought is usually a disaster impacting a wider region, the entire village will see its water resources diminished. In this case, another village from the neighbouring region which is not impacted by the water shortages can step up to transport water to our village. This will however pose a relatively large burden on this second village. Transporting something as essential as water requires adequate infrastructure, be it pipe networks or roads to transport water by truck. This type of larger investment of resources might be next to impossible for one or two villages, but would be relatively easy for cities with a much larger population. And when disasters become even larger, a neighbouring region or even country can fulfil the role of the object to interact with our impacted person, community, village and region.

But interactions between people, communities, regions and societies are not limited to insurance in times of hardship. While these interactions do indeed provide a lifeline when needs cannot be met, they also provide wants. It is well-known that no two people are alike. Every individual person brings something unique to their community, every community brings something unique to their region, and every region brings something unique to the country. Some people, communities, regions and societies might be more skilled in agriculture, while others are more focussed on heavy industry, research, arts, shipbuilding or any of the many facets of human life. And while every community might have their own baker, some regions may be well-known for the quality of their bakeries.

The point to make here is that interaction makes life easier. For instance, while I might be adept at using a sewing machine, someone else will be highly skilled at it and can produce the same amount of clothes faster and with better quality. Some regions have highly productive farmlands due to geographical conditions but do not possess a lot of mineral deposits, while others have rich mineral deposits but as a result lack substantial farmland.

Let's take the cases of two countries, the Netherlands and Russia. Although they both possess farmland of similar quality, they still might not be able to produce the same crops with equal efficiency due to certain crops being endemic to a certain region. Take for instance the simple cherry. The Netherlands possesses the sweet variant of *Prunus avium*, whereas Russia sees *Prunus cerasus* as its endemic species. And while each might be able to grow in the other region, they tend to thrive in their endemic zones.

An interaction between the two cherry-producing countries might not be necessary, as both can produce their own fruit. But it can still be beneficial. Both countries can learn from each other how to take care of their own cherry trees, both can learn new recipes and enhance their cuisine with flavours otherwise not available. And let us not forget the interactions between coupled objects. The interaction between Russia and the Netherlands in terms of cherries might impact many other coupled objects in the area of trade or international relations.

By seeing individual people – but also communities, regions and larger societies, as well as the interactions between them – as objects we can obtain our second rule of sustainability.

(2)

An object, of whatever form, should be allowed the freedom to pursue its wants and needs in ways it sees fit, whilst fulfilling its obligations to objects to which it is coupled.

The above rule follows on from the first rule of sustainability as discussed before. Every object is inherently engaged in interactions with other objects. A person cannot exist without the communities they live in; the communities cannot exist without the regions and countries they reside in. And vice versa, countries cannot exist without the communities and people residing within them.

This inherent and constant interaction between these various levels of human civilisation brings benefits to each participant, but also responsibilities. The benefits can take many forms – financial, intellectual, social and otherwise. As for the responsibilities, these will be elaborated on in the following rules of sustainability.

2

On the interplay between progression and conservation

Ability to learn

All species in existence today pose a certain capacity to learn. Learning is synonymous with the ability to adapt. Take the example of food: I have always eaten wheat products such as bread and pasta. Now there comes a pest that decimates the entire wheat harvest and will continue to do so every time I try to replant the field. There is no reason for me to perish at this moment, because there is plenty of other food out there. But if my entire knowledge and skill set have been focussed on cultivating wheat, baking bread and making pasta, I will likely lack the skills to forage for mushrooms, to hunt for deer or even to cultivate another crop. In order to switch to another food source, I will need to learn how to use it, either through somebody teaching me how to hunt, or explaining to me which mushrooms are edible. I might simply try different foods and find ways of cultivating and preparing them by trial and error. Even when I would not learn by the modern conventional means of books, schools, evening classes and podcasts, the simple fact of my continued existence proves that I have learned how to use a new source of food. Without learning, I would indeed have perished.

This ability to adapt holds true for all creatures. Many species have been observed to use tools in a multitude of ways and they teach these skills to their offspring. Crows for instance have learned to drop nuts on a road with a traffic light, let cars crush the hard shell of the nut and wait for the traffic signal before eating their spoils.[7] This behaviour obviously did not exist before humans invented the automobile.[8]

There are less obvious behaviours which can constitute as learning. A mushroom that grows solely on the decaying matter of a specific plant might experience trouble if that plant species suddenly disappears. But many species have already learned over the course of evolution that using a single source of food is unfavourable. Our mushroom might prefer a very specific plant, but can very well survive on the remains of other plant species.

There are however still cases of highly specialised species, meaning that they depend on a single source for their wants and

---

[7] Crows using cars as nutcrackers has been studied for at least seventy years. Some have disputed the behaviour being intentional, to which I disagree. All reported observations denote a similar behaviour – The crow places a hard-shelled foodstuff in a lane of traffic, waits patiently nearby, observes the food and only flies to it and starts eating it after it has been crushed by a car. You can search for instance for a video of a crow in Akita, Japan and be the judge whether it was intentional or not.

[8] I am not aware of reports of crows using horse-drawn vehicles to crack nuts.

needs. Typically, but not exclusively, found on remote islands or in difficult to navigate terrain, these species might very well be thriving because of their specialisation. An example is the koala in Australia. Their region is a home with many eucalyptus trees providing ample food. Though due to the increase in global temperatures and the subsequent shift in ecological biomes, these eucalyptus trees might see a future where they are not be able to grow as well, or even at all.[9] In this case the koala will likely face a similar fate.

But the hypothetical extinction of the eucalyptus and koala partnership does not disprove the ability of either species to adapt. Rather, present-day climate change is unfolding at a speed that could very well outpace the learning ability of both species. Creatures learn at different rates. And learning rates also depend on what needs to be learnt. For humans it is rather easy to learn directions to a new supermarket. But if all available supermarkets for whatever reason exclusively sell eucalyptus-based products, the human body will likely not learn quickly enough to make do with the new nutrient-deficient diet.

Besides species learning about what they can and cannot eat, where to find water and which materials and tools they can

---

[9] A biome is a geographical region with its own specific climate, flora and fauna, and nutrient flows. The location of such a region can change over time due to for instance a changing climate. The plants and animals within the biome then migrate to regions which maintain their preferred climate conditions. But every species has its own speed at which it can migrate, large animals often faster than smaller animals and plants. And although eucalyptus trees are known to grow fast relative to other trees and as such might be able to migrate relatively fast as well, the Australian wildfires of 2019-20 (ca. 18 million hectares burnt) and of 1974-75 (ca. 117 million hectares burnt) show that changes to a biome can still have serious impacts even to the fastest moving species.

use to make shelters and protect themselves, there is another phenomenon that many species learn: societal organisation. Take for instance a bee colony with its classes of workers, drones, guards and its queen. The amount of soldier bees is shown to depend on the presence of predators, such as wasps, in the area.[10] If you were to place a beehive in an area devoid of predators or remove these predators from their surroundings, the number of guards will decrease over successive seasons as the hive can divert more of its resources to the collection of pollen. Additionally, the destiny of the queen depends on many factors. If she stops producing larvae, she will likely be exiled or killed by the colony. If temperatures drop, the colony will bundle closer together; if temperatures increase, they are more inclined to fly out and explore. Simply put, they behave according to external stimuli.

A study conducted by H.S.G. Dona and her colleagues in 2022 showed that bumblebees enjoy playing with small balls.[11] This is obviously not typical behaviour, but under laboratory conditions the bumblebees learned quickly that they could move the balls and seemingly derived joy from doing so. To me this shows that learning does not necessarily need to be tied to needs alone,

---

[10] Allen Moore et al. described in the 1987 paper *'The guard honey bee: ontogeny and behavioural variability of workers performing a specialized task'* how worker bees temporarily took up guarding as a task, where the amount of guards differed between colonies. Almost 40 years later, in the paper *'Division of labour during honeybee colony defence: poetic and scientific views'*, Daniela Ramirez-Moreno et all. reviewed how guard bees recruit soldiers when a threat is detected. The amount of bees recruited for "military service" is determined based on the perceived threat, among other factors.

[11] If you enjoy high quality research as well as watching bees have fun, please do yourself a favour and read the paper *'Do bumble bees play?'* by Hiruni Samadi Galpayage Dona et al (2022).

but can also occur in terms of wants. If these bumblebees show the ability to learn that the balls in the lab can be played with, perhaps the ability to play can be found under natural conditions with objects and in ways that we humans simply do not yet understand.

Bringing it back to OOO, we might want to elaborate on the question of whether all objects have the ability to learn. We may again observe the humble rock. We discussed the wants and needs of the rock in ways similar to living objects. But does this imply that the rock has the ability to learn?

Frankly, I have put an insanely large amount of time in pondering this question.[12] My first thought was a rather straightforward 'no'. A rock does not possess the ability to learn and adapt to its surroundings; it will simply go with the flow. Even though it has a need to not erode into oblivion, as mentioned before, it does not seem to undertake any effort to counter any eroding effects. It will not hold on to something to stay out of a raging water cascade. For sure, the waterfall might have a nook in which our rock gets stuck, preventing it from going further down the water system. But it is the waterfall in this case that operates as the acting object, not our rock that simply accidentally got stuck. Our rock will also not develop a tougher exterior to withstand the forces trying to erode it. Or will it?

Anyone who has spent time in or around upstream rivers can testify that rocks occupying the raging waters are rounded and smooth. Although this is typically understood as the result of erosion, to me it points to something more. A rounded rock is much less prone to further erosion. Either because no protruding edges can get hit by other rocks, which would cause

---

[12] Or is it a large amount of time pondering an insane question?

chipping and degradation of our rock. Or by the aerodynamic forces of the water that flow more smoothly around the rock, leading to less damage from the water directly and less chance of being swept further down the water system. To further explore the learning capacities of the rock, let us investigate within a geological timeframe. Rock, much like everything else on this planet, exists on a cyclical path. Starting with the molten magma stage deep underground, after time the magma will find a path to the surface where it solidifies into stony material. From this stone, rocks will erode and detach. And through many different pathways, and in spite of our rocks' best attempts to slow the process, they will eventually find their way to places where they are forced underground through tectonic movements and will again become magma.

This cyclical pathway typically occurs over many millions, if not billions, of years. At the end of a cycle and the start of a new one, magma is again forced to the surface to become a new rock. Either in the vicinity of subduction zones in the form of volcanic activity or as a mid-ocean ridge.[13] Volcanoes often arise with relatively sharp peaks that are subsequently more prone to erosion. Conversely, mid-ocean ridges are stretched out over vast areas, allowing magma to rise to the surface in a blunter shape which is less prone to erosion.

Here we have an interesting dichotomy between the wants and needs of the various forms of rocky material; magma, solid

[13] When two tectonic plates collide, the heavier plate 'dives' underneath the other which forces it down into the Earth where it eventually melts. These subduction zones are typically characterised by earthquakes and volcanic activity, of which the Pacific Ring of Fire is a famous example. Mid-ocean ridges are conversely created by diverging tectonic plates. This allows magma to flow to the surface and as such fill the space between the plates.

rock masses and loose stones. The magma needs to remain at a sufficient temperature to maintain a molten state and wants to stay in motion. The subduction-zone volcanic activity is thus more beneficial for magma as it allows less area to be exposed to the colder conditions at the surface. The solid rock mass arising from magma reaching the surface needs to grow quicker than that it erodes and wants to erode as little as possible. And the loose rock needs to break free from the larger mass while resisting further erosion and wants to remain as large as possible.

These wants and needs might appear to contradict each other at first glance. But let us consider the interaction between the wants and needs of magma, mountains and rocks as a coupled object. This coupled object needs to maintain all three forms of stone. If the mountain were to not erode, no rocks would form that will eventually find their way to subduction zones to be recycled into magma. In that case, the mountains would grow until eternity to a point where all magma has solidified at the surface of the Earth, assuming a hypothetical circumstance where this would be possible. If all magma were to solidify, we would no longer consider it as its own object where instead it has turned into part of the all-encompassing rocky mass that is the Earth-mountain. The mountains would be the only object remaining at the expense of magma and rocks, and there would be no cycle.

Conversely, if magma were never to find its way to the surface, no mountains and ocean ridges would form. Consequently, no rocks could then form. Again, the cycle would come to a halt with magma being the only object arising at the expense of mountains and rocks. Finally, if mountains and ridges were still to form and the erosion of these objects allowed to continue, but the resulting rocks unable to return to the deep underground, the system would again halt. This time, only rocks – likely in the form of sand – will remain without mountains and magma.[14]

The triad of magma, mountain and rock needs to maintain a balance between each of the three for its continued existence. Even though each one can exist without the other two, the coupled object with its own wants and needs, will demand the individual objects to provide to its wants and needs even when it demands a small sacrifice from the wants and needs of the individuals.

This thought experiment shows a learning ability from the coupled system. In a distant past, in its early stages, planet Earth was one massive sphere of magma. Over time this cooled down to a point where the surface became solid. But the layer underneath remains liquid to the present day. Instead of slowly solidifying completely over time, the rocky material that makes up the Earth has found a balance between solid and liquid. And it appears that this balance is maintained by allowing the exchange between the two.

There are two ways for this balance to be shifted: First, the Earth could receive less energy from within due to for instance

---

[14] Sand in this case might also be seen as a distinct object. Sand (and even smaller particles such as clay and silt) typically originates from the weathering of rock masses into increasingly smaller stones. When such a stone breaks up into millions of tiny fragments of the size of a grain of sand, the collective of these grains behaves differently than the original rock. So in this case, the object of 'rock' will make way for the object of 'sand'. Sand can however erode further into smaller particles, or be compacted deep underground into sandstone. For sand to continue existing, a continuous supply of new particles is thus required. Hence sand might better be considered part of the coupled object of magma, mountain and rock as a part of the cyclical path of stony material.

a reduced radioactive decay of isotopes, which will allow more solid matter to be formed at the expensive of the liquid; Second, the Earth could receive more energy from somewhere else, for example an asteroid collision, leading to more liquid matter over solid. Since these are geological time frames, it is impossible to predict the future accurately but a likely scenario is not beyond the carrying capacity of the coupled object. The coupled object in this case would eventually return back its original state, albeit in a reduced form. It would have learned that, based on the circumstances in which it exists, the triad of magma, mountain and rock is the preferred form. And within this triad each form will have learned about the optimum in which to maintain its existence.

To thus answer the question if a rock possesses the ability to learn, my answer would be yes, but within the confines posed by the coupled object it resides in. If it were to go beyond these confines it would receive pushback from other objects, such as the magma and the mountain, but also from other objects, for example the upstream fish that nestle in the rocks.

The statement that learning means adapting to external circumstances can be understood in terms of balance, and can be summarised in the next rule of sustainability.

(3)

An object, of whatever form, is capable of learning, meaning the forming of an optimum of its own wants and needs, adapted to the confines shaped by the wants and needs of objects to which it is coupled.

An object's learning can be thought of as a crowd of people entering a music venue. Some people prefer to stand as close as possible to the stage, while others aim to stand farther away to get a better view or to maintain some distance from the loudspeakers. Upon entering, the first people pick their preferred spots. But when more and more people enter, more pushing ensues. Someone might be trying to find a new spot, because a tall person came to stand in front of them. And midway through the concert some people start moving to the bar. The exact position of each individual in the hall is dynamic throughout, impacted by their wants and needs as well as those of the surrounding people. The total system is bound by the concert hall. Although no fixed standing spots exist, people eventually find the spot that best fits their wants and needs within the constraints posed by the other visitors. Although one's final spot might not be ideal, they might still find part of the stage blocked by the head of the person in front, but they will enjoy the concert.

New and shiny

The type of learning described in the previous section is something undertaken by the object itself. It enters a new space or its old space changes and it has to learn how to fulfil its wants and needs. Learning however also comes in a different form, something not only performed by the object in question but rather by all the objects before it. In the case of humans this manifests as societal or cultural behaviour. This is learned by a prior object and passed on to the subsequent object, like from a parent to a child, allowing the latter object to learn without having to directly experience the constraints and dynamics of its surroundings itself. And let us call it 'inherited behaviour' for simplicity.

Inherited behaviour can be very useful for an object. For example, it might instil a roster of which foods are good to eat and which are toxic, without having to figure it out by itself with all the associated risks. Or it might provide a framework on how to interact with other objects such that conflict can be avoided or a bond can be strengthened. Overall, behaviour can significantly improve an object's ability to fulfil its wants and needs, while reducing harmful risks.

But inherited behaviour can also stifle further learning. It could inhibit an object's search for different approaches to familiar challenges. Essentially, inherited behaviour leads to the distinction between progression and conservation. A simple example to highlight the interaction between inherited behaviour, the fulfilment of an object's wants and needs, and possible conflicts this could lead to, can be found in birds that eat worms. Let us imagine a common blackbird, a species frequenting my yard and a pleasant distraction to observe from behind my desk, that likes to rummage through the fallen leaves in its search for insects and worms. It usually spends five to ten minutes moving from one side of the garden to the other, before disappearing to what I presume is the neighbouring building to browse further. Judging by its frequent visits, I assume he is quite content in this area. Now let us assume the hypothetical situation of a severe flood. If my area were to flood for a prolonged period of time, what would happen to our blackbird? The ground-dwelling insects will drown, the worms might be able to survive submerged for some time but are unreachable for the blackbird. I am convinced that our blackbird, a species known for their omnivore diet and adaptable personality, will either fly to another area which is not flooded or will switch its diet to berries from bushes and trees which enjoy the submerged state. Whatever our blackbird will consider as a solution to the changed confines of its old reality, he will simply continue to try to find insects in the leaf litter and not starve because he cannot find them.

Now let us consider the same flooded landscape from the perspective of a jumping spider. *Pseudeuophrys lanigera*, the 'home' or 'domestic' jumping spider, is another species that frequents my garden and also my living room. It is fun to observe it walking on the walls of the room, from the plants in the window sill, to finally find a meal under the ventilation grill above the window. They are fun to play with too. When one walks across my desk, I often scratch the desk some distance behind him, which leads to him quickly turning around to observe what the noise was, only to then follow my hand to my face and rotate its head like a dog trying to understand the confusing behaviour from this giant. This hopefully helps to paint a picture of the jumping spider as an intelligent creature, on par with yet very different from our blackbird. What would happen to our jumping spider in the case of a flood? I think that it will have a much harder time surviving the new circumstances it has found itself in. It might find an object to climb to evade the rising water, but only when such object is within reach. As jumping spiders have a body size in the order of millimetres, traversing large distances takes them a relatively long time. Add to this the likely fact that the start of the flood will merely look like an average rain shower. Such a small creature will shelter to try to avoid the large rain drops, but will probably not recognise a flood as such until it finds himself completely surrounded by water. If an object of sufficient height is then within reach, the spider will climb it. But failing to recognise the flood as something different than a normal rain shower could very well lead to it being stranded without an upward escape. Giving some credit however to the spider that does manage to escape to a higher level, I am convinced that it will manage to survive. Jumping spiders are not picky eaters as long as their prey is within a certain size, and ours would likely survive happily on the waterborne insects that thrive in a flooded landscape.

This example might be somewhat unfair to the jumping spider, because its size very well comes into play here. But it does highlight that the spider and the blackbird have different needs in the event of a flood. A bird does not need to recognise the flood early, because he can rather easily move the large distance to dry land. But observing the water as a flood early on is paramount to our spider, as he cannot traverse the required distance in time.

An individual jumping spider with inherited behaviour allowing it to observe a flood early on, will have a higher chance of survival than its peers. If this were told in a children's story, the form will likely look something like this. It starts to rain and puddles start to form on the ground. All the jumping spiders climb in the crack of a small wall to shelter from the water, aside from one individual who climbs up a tree. The spiders in the wall make fun of the spider in the tree for being stupid, as he will be at risk of being eaten by birds. But our one spider knows that birds do not fly with such heavy rain. And when the water starts to rise higher and higher, the wall spiders scramble to find higher ground quickly. In their rush, some perish. And when the flood disappears, the surviving spiders teach their offspring about the tree spider.

This fairytale highlights inherited behaviour. One individual dares do something different which turns out to be the better way, after which the rest of the community start mimicking the behaviour. Another such example can be found with the Eurasian jackdaw. They often congregate in vast numbers, seemingly engaged in some sort of congress in which they discuss jackdaw business. While no human can truly know what they discuss, if anything, it is known that jackdaws observe each other thoroughly. For instance, if one individual appears to be eating well it can trigger a change in social associations within a group.[15]

The positive feedback loop of one individual daring to deviate from typical behaviour leading to new inherited behaviour in a community knows many more examples. But a negative feedback loop, or better yet an inhibiting feedback loop, also exists in which no individuals dare to take a leap away from the inherited behaviour even when circumstances might demand so. Take the example of the golden toad, a now extinct species that lived in a relatively small area within the rainforest in Costa Rica. It was declared extinct in 2004 by the International Union for Conservation of Nature (IUCN) after not having been sighted since 1989. Although the exact cause of its extinction is not formally known, with hypotheses ranging from climate change to a fungal infection, we can use the climate hypothesis for our example.[16] It states that rain was absent and temperatures were high due to the El Niño event and possibly also due to human-induced climate change. The toads needed moderate

---

[15] In the 2023 paper '*Wild jackdaws can selectively adjust their social associations while preserving valuable long-term relationships*', Michael Kings et al. conducted a compelling study into the social learning of jackdaws. Although social associations appeared changeable based on an individual's ability to acquire food, long-term relationships such as family ties remained stable.

[16] The hypotheses mentioned are taken from Tim Flannery's '*A warning from the Golden Toad*'

temperatures and sufficient water in order to mate. This prolonged period without rain and with temperatures that were too high resulted in the toads deeming the conditions of their surroundings to be unfavourable to raise offspring. It's likely that the toads had inherited behaviour that told them that their offspring had a better chance of survival if temperatures were low and water were ample. The conditions resulting from El Niño could very well have been engrained in their inherited behaviour as the event occurs cyclically. Perhaps the toad's coping mechanism was to wait for the event to end, before resuming their business-as-usual approach. But human-induced climate change might have increased the severity of El Niño to the point where the toads simply had to wait too long, weakening their population with each event to the point of extinction. Deviating from their inherited behaviour would thus have been crucial to their survival as a species. Had some toads decided to have offspring regardless of the environmental conditions and made do with the limited resources that they had during a time of drought, the others might also have learned that less water was needed for their offspring to survive than their inherited behaviour told them.

This example taps into the rate of change of inherited behaviour an object can attain. Here an important distinction can be made between individual objects and collective objects. A capacity for higher rate of change can be observed in individual objects than in their collective counterparts. We can look at Baltic herring as an example. This species of fish, like all herring species, can theoretically reproduce at any given time of the year. Yet we know that sometime around the mid-sixteenth century, the population of Baltic herring that reproduces in autumn collapsed due to a combination of a colder climate and overfishing by humans.[17] Only since this collapse has the autumn-spawning herring been replaced by a spring-spawning population. Naturally it appears that the collective object of Baltic herring had learned that is more beneficial to reproduce in spring, but not before enduring a dramatic population collapse. Here it is also important to realise that it was a limited number of individual herring that quickly switched to spawning in spring, as without them the complete subspecies would have gone extinct. The collective object thus learned, albeit slowly, from individual objects to change its inherited behaviour.

Since the individual and collective objects thus contain inherited behaviour that can change with a certain rate of change, let us investigate this phenomenon for the coupled object. Here I want to include another element as well, which is that of inherited behaviour that only indirectly impacts the wants and needs of the coupled object as well as the objects that it couples. The coupled object I want to examine here is that of the human understanding, or rather misunderstanding, of the composition of wolf packs. This misunderstanding might not directly impact either species. After all, a wolf and a human will not harm each other for not understanding the other's group composition. Indirect impacts, however, are plenty. A lack of understanding from the side of the human can, and does, easily spiral into a hostile attitude toward wolf packs, often leading to violent conflicts and deceased wolves. Additionally, humans have often misused the false notion of wolf packs containing an alpha male to project a hierarchy onto society where certain men, or men in

[17] As per the study by Oliver Lehmann et al. titled 'De vorbisteringhe deß heringheß uthe deme Sunde Indizien für eine vorindustrielle Überfischung des Ostseeherings' (2021).

general, are supposed to be positioned above others. For clarity, a wolf pack is nothing more than a family, very similar to their human counterpart. A pack consists simply of a mother and a father wolf, perhaps same-sex relations exist, and their pups. Once the pups become old enough, they 'move out' to build a life and a family of their own.

The misunderstanding of the family dynamic of wolf packs dates back at least several decades to research conducted on captive wolves that were placed in an area with other unrelated wolves.[18] This yielded results that were falsely projected onto the whole species. Yet decades later, a large portion of human society still believes these false results. Although several individual objects, humans, have since adopted this new knowledge, the collective object of human society still stubbornly follows the old false rhetoric. Again the inherited behaviour of the collective object shows a slower rate of change to that of the individual object.

This also holds true when observing the aforementioned coupled object from the perspective of the wolves. A wolf might not know humans having the same type of family dynamic as their own, that of parents and children living together until the children reach adulthood. By itself this lack of understanding is unlikely to cause direct consequences. But in the hypothetical scenario where a wolf mistakes a small child as an easy prey, it will inevitably evoke the wrath of the parents and likely that of the entire community or society. Because of this wolves

[18] Although David Mech's book '*The wolf: ecology and behaviour of an endangered species*' (1970) was the main culprit of the misunderstanding of wolves, he spend many years fighting the persistent myth that he helped to create. His book '*Wolves: behaviour, ecology, and conservation*' (2003) is probably his best attempt at this.

have learned over the course of many generations that humans are to be feared and avoided, or persecution will ensue. But these conflicts have become somewhat more prominent in recent years, as wolves have reintroduced themselves across Europe. They have grown in number from small populations that have survived due to being isolated from humans. These new generations are less careful when it comes to interacting with the humans that they share land and resources with. Even though since recently many wolf families have been slaughtered in Europe on the basis of false rhetoric, the remaining wolf packs are not always as careful and avoiding as they might need to be to avoid further conflict.[19] Over time they will likely learn again, but it remains to be seen how much time this might take.

What I am trying to highlight with these examples of inherited behaviour is that it can be very useful in providing safety and success for an object on the one hand, while also limiting the ability to adapt to changing circumstances on the other. Inherited behaviour is an important factor to determine the rate of change an object can achieve, influencing whether an object can adapt to a changed reality quickly enough or if it will perish. An object that cannot maintain a rate of change dictated by its surroundings will face severe adversity. But similarly, the object will face negative consequences if it exceeds the required rate

[19] Since the delisting of wolves in 2024 from "strictly protected" to "protected" under the Bern Convention, several European countries have been systematically eradicating part of their wolf population. And it is Switzerland that ironically seems to be leading the culling. This culling is typically targeted towards entire packs. But in practise it has been proven difficult to do so. Add to that the occasional killing of a single wolf which has attacked livestock and it slowly becoming clear that entire wolf families are being torn apart. What consequences of this will be on wolf dynamics and the behaviour to avoid humans, remains to be seen.

of change.

Let us dive into the case of *Homo sapiens* over the course of the past 12,000 years or so as its own object, after first investigating some less obvious objects. We can start with the coupled object of the Oostvaardersplassen, an artificial nature reserve created in one of the Netherlands' newest polders. What started as a project to create as natural a wetland area as possible grew into a nature reserve complete with countless bird species as well as large grazing herbivores like Konik horses, Heck cattle and red deer. They forgot to add large carnivores but we will discuss that in a moment.

We shall observe the Oostvaardersplassen as a coupled object, rather than an individual object (a singular nature reserve) or as a collective object (the sum of all species that make up the nature reserve). Depending on the context, one might also be interested to investigate this area as one of the other object types. However, to see it as a coupled object allows us to examine the interplay between its intended purpose and its observable outcome, created by the various species that inhabit the area. Briefly put, the Oostvaardersplassen was never originally meant to be a nature reserve and instead was built as an expansion to the Dutch agricultural land. But during the many years of constructing such a large polder, a growing number of citizens pleaded for the continuation of the new nature that had begun to establish itself on the former seabed. And so it was decided that, since any polder is essentially a blank canvas for policymakers, the area was to be built according to a different vision – a reserve that was maintained by nature itself.

What nature had started by itself after the poldering process was complete, however, was the establishment of pioneer plants, some insects and mostly birds. And then policymakers decided to install a small herd of large grazers in the area.[20] And it should come to no surprise that these grazers had the time of their lives. Because as previously mentioned, people conveniently forgot that these grazers' main predator, the wolf, had been eradicated from the Netherlands many decades prior. So without any form of predation, these grazers were grazing ever so happily, and multiplying year after year. Their number became so great that even a slightly harsh winter resulted in the starvation of dozens of animals which in turn sparked campaigns of the general public in favour of feeding the animals in the reserve. Ultimately the management of the reserve decided on a maximum number of large grazers in the area which are now fed if they show signs of starvation, and are culled if they become too many.

Now going back to the coupled object, the topic of our discussion at hand, we see that the intended purpose of the area did not match the observed outcome. The large grazers adapted too well to their new surroundings. In turn, the inherited behaviour of the coupled object showed a rate of change which exceeded that of its surroundings. Allow me to elaborate. The reserve was intended to provide a place for nature, safe from human activity. And that it did, rather spectacularly. The issue was with the type of nature that developed, which did not meet the expected outcome. The nature that developed so spectacularly, but which was not intended to develop, shows run-away development that overshoots the wants and needs of its surroundings and hence that of itself.

It did so firstly on the count of the inherited behaviour of the

[20] For those of you who have ever played video games like zoo tycoon, it was essentially a grab- and- plonk of animals. "There you go, have fun!"

grazers, which tend to multiply more when they feel safe and have an abundant source of food. Of course, the grazers did not change their behaviour of multiplying after large numbers died of starvation because each spring showed a new abundance of food. But this behaviour of grazers is not in itself a sign of an attained rate of change that exceeds that of its surroundings. Quite the contrary, they did not sufficiently adapt to the reality of their surroundings until this very day, which shows a rate of change seemingly lower than that of its surroundings.

Secondly, these grazers in turn impacted the nature reserve by profoundly changing the vegetation. The reeds of the marsh areas have largely made way for grasses due to the pressures of grazing. Here both the reeds as well as the grasses have shown a rate of change in behaviour exactly in tune with their surroundings. The grasses have colonised larger territory on the count that the changed reality (i.e. the introduction of the grazers) diminished their floral competition. The reeds have not completely disappeared because they found areas in the reserve where grazing pressure is still sufficiently low.

Thirdly, the bird species that frequent the area have changed significantly as a result of the changes in vegetation. Some have decreased in population size while others have increased. Both logically followed the decrease or increase of their respective floral needs.

There is also a fourth development although less often mentioned in the case of the Oostvaardersplassen, which is that of fish. The reserve is in part built as a wetland. But the fish that inhabit these swamps are few, both in number of species and of individuals. Add to this the overlooked fact of the complete decimation of the former shallow sea ecosystem for the sake of the new polder, and it appears that the fish have not shown a rate of change in their inherited behaviour that could maintain the change of their surroundings. And as a result, fish have largely disappeared from the area.

These individual (or collective, depending on your frame of interest) objects show inherited behaviours that either neatly follow the rhythm of their surroundings, allowing them to maintain their own wants and needs in a changing reality, or inherited behaviours that fall short of that rhythm, leading to suffering and decay.

But when finally observing the nature reserve as the sum of its constituent species with the intention of being a virtually all-natural area, and the actual reality of the remaining human-maintained area, we see an inherited behaviour with a rate of change that exceeds that of its surrounding reality. And a similar suffering and decay as a result of this excess occurs as would be the case when the rate of change was insufficient.

Here it is important to note that any area of nature, strives inherently for balance. But the achieved equilibrium can be one of high or low biodiversity. Had the Oostvaardersplassen be a fully natural area, meaning an area of sufficient size with predators, it would likely achieve a complex mixture of wetlands, grasslands and woodlands, and all the associated floral and faunal species. Instead, the biodiversity of the area as it stands today, without predators, is lower with mostly grasslands species. The nature reserve, given the circumstances it operates in, is thus still following its inherited behaviour of developing towards a state of equilibrium. But this behaviour will lead it to a state of mostly grassland with large grazers following a boom-bust cycle every year and all other species facing severe pressure.

A coupled object that overshoots the rate of change of its

surroundings will thus result in a less-than-desired state for the object. Although its needs might still be attained (i.e. a state of equilibrium), its wants will be adversely impacted (i.e. low instead of high biodiversity).

If we now observe the interactions between the history of human agricultural civilisation, roughly the past 12,000 years, and nature as a coupled object, the same conclusions will be found. Again we take as given that nature strives towards a state of equilibrium. Likewise, human agricultural civilisation also strives for an equilibrium – that of a secure, stable and predictable harvest year after year. These two equilibria do not necessarily contradict each other. As a matter of fact, they might even be considered one and the same equilibrium under ideal circumstances. If humans are considered part of nature (which many of the environmental movement rightly argue they are), the mere existence of human agricultural workers in nature could yield an equilibrium much alike to the presence of wolves, deer, mushrooms or any other object.

Of course, the real world differs significantly from the ideal situation. Somewhere in its striving for an equilibrium, human agricultural civilisation has superseded with the rate of change of their inherited behaviour that of their surroundings. And as a result, the coupled object of the equilibrium between human agricultural civilisation and nature has suffered greatly.

Human agriculture developed organically from the previously dominating hunter-gatherer cultures. It appears only logical that humans at a certain point in time learned that plants grow from seeds and that humans could physically move these seeds across the land. For obvious reasons humans would plant these seeds close together, rather than having them spring up scattered across the land which would have a person have to search far and wide for the plant. Having edible plants closer together and knowing exactly where these plants are, apparently proved to be a huge benefit to the human species as agriculture became the norm for most of humanity in the following millennia.[21] In itself this can very well be part of the natural equilibrium. After all, many species – such as birds, mammals and of course plants –disperse seeds in a more or less predictable pattern.

Problems arose when humans began to nurture plants in the places where they had dispersed the seeds, especially when the location of the new seed differed significantly from the more natural locations where similar plants would grow. For a seed to germinate, it requires manipulation of its direct surroundings so that it can root properly and that no undesired plants outcompete the seedling. This behaviour, for example tilling the soil or removing surrounding plants, differs greatly from bird-assisted seed dispersal. Birds simply drop a seed at more or less predictable locations, for example near nesting habitat, but will not tend to the resulting seedling. This allows the equilibrium of nature to decide whether a seed will grow to maturity or that other species will outcompete it.

But human agricultural society requires an equilibrium state

---

[21] There are certainly societies that still rely on hunter-gatherer tactics, such as indigenous communities within the arctic circle hunting whales. And one might argue that pastorally nomadic people such as the Bedouin are practising a form of hunting-gathering based on their continuous search for suitable land to bring their animals to. There certainly is also scientific research available which indicates that agriculture has actually been hugely disadvantageous, although this might heavily depend on which region and hence which climate is being assessed. Following the observation that agriculture has been widespread for millennia, this to me signifies that societies throughout history apparently saw benefits to such a system.

to acquire a stable and predictable harvest. By simply relying on seed dispersal akin to birds, one could never be certain of a harvest. The human agricultural object thus became dependent of manipulating the nature object around their seedlings. This is the first point of friction between the two equilibria within our coupled object.

However, the rate of change of the inherited behaviour of human agricultural society superseded that of nature in another way. The speed with which humans became able to experiment with manipulating the surroundings of seedlings began to increase. No longer were human settlements solely clustering their plants closer together and manipulating their direct surroundings, they were now also bringing in water from farther away through for instance irrigation systems or removing excess water through drainage channels if it showed that it made their plants of interest fare better.[22] They were removing trees to allow for more sunlight to increase growth. And they began bringing in seeds of plants which were previously far outside their reach as hunter-gatherers, such as exotic plants from regions far away.[23]

With every change made within the human object, the nature object began struggling more and more to attain a new equilibrium, its principle want. After all, how could it attain an equilibrium after a change in its surroundings if it did not even manage to reach a steady-state after the previous change?

Now we could argue that the individual object of the natural

---

[22] I am referring to developments in agricultural practises in for instance the Fertile Crescent several millennia ago.

[23] Many examples can be found for humans hauling plants across the world; from Polynesians bringing plants to remote islands, to the spice trade from European colonists.

equilibrium simply did not achieve a rate of change in its inherited behaviour that allowed it to follow the rate of change of its surroundings, whereas the individual object of the human agricultural equilibrium did achieve this. And that this would mean that the former individual object simply struggled to meet its wants and needs, while the latter individual object flourished. This would, however, disregard the coupled object that exists between these two individual objects. Much like the discussion of the triad of rock, mountain and magma previously, human agricultural civilisation and nature must allow each other their own wants and needs to a certain extent to maintain those of their own. If human agriculture would completely obliterate nature to achieve its own wants and needs, it would find that water would become polluted, soils poisonous and storms frequent. This is, of course, a gross simplification and I will go into more depth in chapter 3 to do justice to this highly important topic. The result of denying the individual object of nature its wants and needs will result eventually in a situation where the object of human agriculture will see a move away from its previously attained equilibrium. This will be due to the human object changing too fast compared to the nature object within the coupled object.

The examples giving in this section highlight the importance of inherited behaviour in the learning process of any object. Furthermore, they show that since every object is inherently linked to others (as per the first rule of sustainability), the inherited behaviour of an object should align to that of the objects it is coupled to, in order to maintain its own wants and needs. This results in our next rule of sustainability.

(4)

Learning, that is the sum of an object's ability to react to its surroundings and the behaviour it inherits from the objects it descents from, should maintain a rate of change to an object's behaviour which is attuned to that of the objects to which it is coupled.

Allow natural progress

Let us define progression as a longing for the new and conservation a desire for the familiar. These two seemingly contradicting terms have more in common than one might expect at first glance. This is especially the case in modern society, where they seemingly have grown to become complete antonyms of each other.[24] But also objects other than *Homo sapiens* and their respective societies experience a constant interaction based on these two terms.

Given the third rule of sustainability, which states that every object is capable of learning, it follows that progression is inevitable if an object exists within a dynamic space. Here, the wants and needs of objects and the object's attempts to attain them are continuously in motion. Given rule 4, which states that an object's ability to react to its surroundings should not exceed the rate of change of the space it exists in and of that of the objects it is coupled to, since this would lead to adverse effects for the object itself. Conservation would thus be inevitable, assuming objects always respond to the adverse effects of their

[24] Referring to the increased gap between the political left and right, at least present in many western countries.

own doing. To prove that progression and conservation are two sides of the same coin, we must thus investigate two hypotheses. First, an object can exist within a static system. Second, an object can indefinitely resist the adverse effects experienced by it having a rate of change in behaviour that exceeds that of its surroundings.

Starting with the first hypothesis, we need to distinguish between dynamic and static systems. A dynamic system is one where all objects that inhabit the system continuously respond to one another in a non-predictable way. We already defined the first rule of sustainability as objects inherently depending on interacting with other objects. But given that in a practical sense there are always changing variables, be it temperatures, water chemistry, landslides, solar activity or quite literally anything else, we know that objects always have to change their behaviour accordingly to the new circumstances it finds itself in.

Even if we could maintain external variables as constant, there will nevertheless exist chaotic (i.e. unpredictable) internal variables. For example, we know that our Sun follows an eleven-year cyclical pattern which affects the type and intensity of the energy we receive from it. It is likely that this is not exactly eleven years, but even if it were exact down to the second, we have a forest of data from decades of research that shows that the amount of energy varies between each cycle.[25] I would argue that this should be considered a chaotic variable. Besides solar

[25] A great source are the datasets produced by the United States National Oceanic and Atmospheric Administration (NOAA) on solar cycle progression and space weather. They show that the peak of F10.7 Flux in August 2024 (ca. 245 solar flux units) was around forty four percent higher than the previous peak in February 2014 (ca. 170 solar flux units). It also shows that these peaks are not spaced exactly eleven years apart.

activity, there is also historical evidence of mass extinction events, such as the Great Oxygenation Event. This atmospheric event began roughly between two and 2.5 billion years ago and continued until today. Current scientific insights point to the evolution of new microorganisms as one of the leading causes of this event.[26] These organisms gained the ability to manipulate nitrogen in a new way, which altered the nitrogen cycles, which in turn altered the carbon cycle leading to more sequestration of stable carbon. This finally led to a massive drop in global temperatures and a surge in atmospheric oxygen, which a significant portion of objects at that time were not equipped for. This sudden (on a geological timescale) change in Earth's atmospheric conditions can be considered as a chaotic variable. And more importantly, the genetic mutations that caused these new microorganisms to come into existence are by themselves a chaotic variable.

Naturally one can make the argument that ultimately every variable is, in fact, predictable. This would suggest a steady system with predetermined changes. But such an argument is based on the assumption that there exists a measurable and predictable universe which can be captured in a set of rules and equations. The pursuit of this – an ultimate reality – is what many people misconstrue as science. But science is rather an infinite cycle of learning what one does not yet know. It is an endless refinement of research questions and hypothe-

[26] There are plenty of resources that describe one or more factors around the Great Oxygenation Event, especially under the similarly named topic provided by ScienceDirect. 'The Great Oxygenation Event as a consequence of ecological dynamics modulated by planetary change' by Jason Olejarz et al. (2021) however is a great study that brings these insights together into one consolidated research.

ses. We are currently investigating the increasingly smaller worlds of quantum mechanics, which behave unexpectedly when compared to our previous understandings. It is already known that atoms are not the smallest building blocks, because they in turn are made up of quarks and muons and whatever smaller particles we may discover in the future. The same holds true for the ever- increasing macro realm. The portion of the universe that we can observe is becoming larger and more detailed each year. Stars and black holes turn out to behave in ways previously unknown to science, for instance through the direct observations of gravitational waves in 2015 by the Laser Interferometer Gravitational-Wave Observatory in the United States and the Virgo interferometer in Italy. But by the time we will have a solid grasp on the exact workings of our own universe, it may turn out that the multiverse theory was correct and science can start over attempting to understand the next universe. And even if we understand the exact workings of the micro worlds, their emergent properties when scaled to the macro realms are typically non-linear. The difference between quantum superposition, where a particle can for instance exist in two places at the same time, and the classical reality of definitive states can be seen as an example for this difference in properties between the micro and macro worlds.

Science might be humanity's futile attempt to make sense of the world around them. But as I have hopefully convinced you by now, this world is dynamic and in no way stationary. Science should thus rather be seen as continuous learning by the human object instead of a finite task, and this learning is unavoidable due to the very nature of the space the object inhabits. The object's space should be regarded as inherently dynamic and no object will ever truly reach an understanding of the absolute fundamentals of its workings. There will always exist random variables to which objects have to respond by learning and adapting their behaviours. We can conclude that this learning and adapting, changes within an object which I call progression, is indeed an inherent pursuit of objects. Thus finalising the analysis of the first hypothesis, which stated that an object can exist within a static system, and disproving it.

Moving to our second hypothesis, we now shall investigate the scenario where an object impacts the objects around it for a prolonged period by *over-adapting* to its surroundings, without it being hindered by the adverse effects caused.[27] In other words, we must look at an object experiencing the absence of a negative feedback loop towards its own behaviour. An example to illustrate this would be the coupled object of climate change and humanity, where the impact humans have on the planet causes all sorts of disasters such as heat waves, floods and storms that direct impact back to humans. As an effect, humans are forced to either change their behaviour or face the consequences.

But a more obscure example, related to warfare, might be better suited in this case. We start by observing war, which I define as the violent interaction between objects in search for a renewed object space of sustainable wants and needs, as a coupled object. This coupled object is ever-present, since objects that are pushed too far away from their wants and needs by the hand of another object will resort to violence if non-violent means of reclaiming its position do not work. Thus

[27] Normally an object adapts to changes in its surroundings, as we discussed as part of the first hypothesis. The term *over-adapting* then refers to the situation where an object changes faster than its surroundings.

we should see warfare as organised violence, whereas violence itself typically exists as an impulsive and intuitive object. I am using warfare and violence to investigate if an object can *over-adapt* to its surroundings to the point where it starts negatively affecting the objects it is coupled with but without experiencing the negative effects which attempt to move the object back in tune with its surroundings.

Warfare and violence have obvious impacts on their surroundings. If for instance two male deer are in conflict over a female, they attack each other to the point where one of them admits defeat and goes away, leaving one male deer without the female. Or if two bears are both claiming dominance over a certain stretch of forest, they might fight each other until one kills the other, resulting in one having to find a new territory. In the case of warfare, one country might be in conflict with another over a certain resource and the war will end if one gains control over it, leaving the other without the benefits of the resource. Simply put, warfare and violence as objects in their own right will always negatively impact some other objects, and by its nature will always diminish. First, violence increases but at some point it will decrease. It might require many objects perishing, but violence will eventually diminish either way. As such it could prove our second hypothesis.

But how about the winner of the conflict, the benefactor of the violence? The winning deer that got the female, the bear that got dominion over the forest, or the state that got the valuable resource? At first it might appear that these objects are not negatively affected by utilising violence. Quite the contrary, they gained precisely what they wanted. But there is a downside to this victory. Female bears are typically very cautious of males as they know them to be dangerous. For male bears this causes significant difficulties in the search for a mate, as females may become violent towards them. Something similar typically happens with the victorious country after warfare. Allies lose the feeling of comfort towards them, and enemies become even more aware of their danger. This typically leads to more isolation and more hardship for the victorious country. And as for the deer, it should be noted the violence between males is due to hormones the females secrete. The violence of the males is thus directed towards each other, leaving the females relatively safe. In addition, the violence exerted by the males is typically not significant enough to cause lasting injuries. Even though they could theoretically cause significant bodily harm by using their antlers on for instance the side of the body, they choose to face each other antler to antler They thus seem to understand that too much violence is not in their best interest, as that would leave them vulnerable. If the negative effects of violence are thus more than the positive, from the perspective of an object using violence as a tool to achieve its wants and needs, violence will be stopped, diminished or avoided.

From my point of view, this disproves our second hypothesis which stated that an object can indefinitely resist the adverse effects experienced by it having a rate of change in behaviour that exceeds that of its surroundings. Violence and warfare might at first cause benefits for specific objects, which would appear to justify more violence to gain even more benefits. But control mechanisms appear to be in place that causes the object of violence to be diminished.

From the combination of our two disproven hypotheses, we can now conclude our next rule of sustainability.

(5)

A combination of progression (learning and adapting to changes in an object's surroundings) and conservation (the avoidance of the negative feedback from those surroundings if an object changes too fast), is used by every object as a means to achieve its wants and needs.

The main takeaway lesson from this rule and the rules that we derived before, is that no object is ever static. Certainly, many objects can be understood as static for practical purposes like cosmic objects such as the movement of our Milky Way galaxy or the decay of our Sun, as their timescale is simply too vast compared to that of the human one. However, objects like the rock and mountain also go far beyond the human timescale yet should also be regarded as dynamic.

3

Cultivating collaboration

Why agriculture?

As of modern times, humanity has become an agricultural species. Although some niches of hunting- gathering still exist, such as deer hunting and mushroom foraging, these activities are predominantly intended for leisure or for high value, exclusive products such as truffles. And although we should not ignore the hunter-gatherer societies that still exist to this day, they represent small numbers compared to the global human population. The vast majority of human sustenance nowadays comes from agricultural practises. Historians and other scientists that focus on past humans have shown us that there was however a time where humans (*Homo sapiens* and their predecessors) were dependent on hunting woolly mammoths and foraging berries for their survival.

The main point here is that there has been a shift in food across human history. Likely this shift occurred gradually over time and space. Many regions across the world saw a quicker or rather slower transition towards the agricultural model, the Fertile Crescent considered to be among the first.[28] What I want to explore here is the link to our previously developed rules of sustainability. To briefly summarise, we concluded that objects are dependent on interaction with other objects for their survival, that each object should be able to pursue its own wants and needs whilst acknowledging that it should not encroach too much on the wants and needs of other objects. We also found that each and every object has the ability to learn how to maximise achieving its own wants and needs whilst ensuring that it does not damage other objects too much, and that it will experience an inherent pushback from those other objects if it does damages them. Finally, we concluded that every object utilises a balance of progression and conservation in achieving the most desirable outcome for itself within the limits set by the other rules.

The transition to agricultural societies could be considered a logical and natural pathway for any object to take given a set of specific circumstances. It is, after all, a path that not only humans have followed. There are species of termites and ants, such as leave-cutter ants, that farm fungi and aphids. There are fish such as the Damselfish that farm algae. The *Littoraria irrorata* snail cultivates fungus. The *Mastigias papua* jellyfish

---

[28] The complete history of modern humans has rather recently been extended to about 300.000 years based on the identification of the fossils from Jebel Irhoud in Morocco as Homo sapiens. This can for instance be found in the 2017 article by Daniel Richter et al. titled '*The age of the hominin fossils from Jebel Irhoud, Morocco, and the origins of the Middle Stone Age*'. Given this timeline, the shift towards agriculture happened rather fast, even if it may have taken several hundred to perhaps some thousands of years.

cultivates algae through a symbiotic relationship. And the Yeti crab, *Kiwa hirsuta*, appears to grow edible bacteria on its arms by living near sulphur-rich geothermal vents. And given this range of species that engage in farming, one might conclude that there must be many more species (and other objects) utilising similar practises that are currently unknown to human science.

I want to take a sidestep here and argue on the further principles of OOO. So far, I have mainly reasoned on the basis that all objects are equal, which OOO explains as every object having its own reality and objectivity, its own wants and needs. The reduction of an object to its own specific set of wants and needs, and stating that those can only be met through interaction with other objects, requires us to avoid the fallacy of under- and overmining the object.[29] As we discussed before, an object can only be ascribed a value based on the context in which it is described. A person, amoeba, stone, the economy or any other imaginable object can either be a collection of parts, or part of a collection. But at the end of the day, a stone is a stone, being more than the qualities another object might ascribe to it, and more than its interactions with other objects.

In addition to the under- and overmining fallacy, we must discuss the withdrawal argument of OOO. It states that an object's reality can never be fully comprehended by another object

---

[29] In his book '*Object-Oriented Ontology, A new theory of everything*' Graham Harman explains the concepts of under- and overmining in chapter 1 (in the 2018 Pelican edition). Briefly paraphrased: Undermining constitutes the reduction of objects to smaller, more fundamental parts such as atoms; Overmining dissolves an object into its effects and relations. According to Harman, undermining fails to account for emergent properties of the fundamental parts of an object, whereas overmining considers an object as nothing more than its relation to other objects and thus denying the autonomy of the object.

and that every object is partially withdrawn, in a metaphorical sense, from every other object.[30] This implies that an object can never completely understand the wants and needs of any other. Naturally, some objects are more comprehensible than others, for instance because of close proximity or similar ethics to an observer. But a one-to-one understanding is unachievable for any imaginable object. Both the withdrawal and the under- and overmining duality discussions from OOO stress that an object remains such in and of itself, regardless of the qualities given to it by another observing object.

This is an important notion in the context of sustainable human nutrition. Humans, like a multitude of other species, depend on consuming other organisms to meet their nutritional and energy demands. Whether vegan, omnivore or carnivore, every person always has to eat other species (be it fungi, plants or animals) since we lack the ability to synthesise energy from sunlight directly. This is an argument often used in works regarding ethics and sustainability, for instance on the concept of the honourable harvest.[31] This concept can be understood to mean that one should never take more than one needs and that certain individuals understand that it is their time to give their energy to another organism.

---

[30] In the first chapter, Harman builds on the under- and overmining arguments to conclude that "there is no direct knowledge of anything". This is further explained as the "quadruple object" in chapter two and beyond, where it is stated that an object can be either real (and inherently withdrawn) or sensual (how it appears to another object, where the amount of withdrawal depends on that object's frame of reference).

[31] An important work detailing the honourable harvest is the book '*The Democracy of Species*' by Robin Wall Kimmerer. It is particularly adept at explaining Native American indigenous knowledge to a Western audience.

But to me the honourable harvest and other writings on sustainable nutrition seem to miss a point. They are attempting to find an ethical standing that allows humans to consume other organisms, which suggests that there exists a way of consuming not currently implemented which is in line with the way of consuming done by all other organisms that require to do so in order to survive. But remember again the first rule of sustainability that we derived: An object, of whatever form, is inherently linked to other objects, and depends on interacting with other objects to provide its own wants and needs. In short, all objects inherently require interaction with other objects for their survival. From here, I want to argue that all forms of interaction lead to degree of competition.

Take an oak tree. It requires sunlight, water and nutrients to survive and thrive. But as any ecologist, biologist, arborist or even an avid gardener will attest, our tree actually requires much more than that. It has a specific ideal range of sunlight and temperature, it requires more of some nutrients, it will perform better when grown alongside a companion plant, and so on. But considering the withdrawal argument, we can never fully grasp what the oak tree really requires. This is more so when considering that no two oak trees are identical. Furthermore, notice how our oak tree requires nutrients and sunlight. By consuming these it denies them from other objects. The tree will have incorporated the nutrients, blocking them from other organisms. And the tree canopy blocks sunlight from reaching the ground, taking away a necessity for smaller plants. But simultaneously, by blocking the direct sunlight, our tree allows for plants that require less sunlight to thrive. These plants would otherwise burn when exposed to the direct sunlight. In addition, our tree provides moisture regulation and local climate formation that leads to more water availability for plants that cannot develop deep root systems. In short, an object can benefit others by limiting some objects.

But would sustainability then be the maximisation of benefits for as many objects as possible at the minimisation of negative effects? I think that this reasoning falls short of the fundamentals of OOO. Ethically it may indeed appear to always be good to strive for the maximum attainable benefits for the maximum attainable number of individuals. But we humans can never fully grasp what constitutes benefits or negative effects to other objects. And even if we could, negative effects will always exist anyway, leading to ethical conundrums such as how much positive is required to offset the negative.

So what does then constitute as sustainable nutrient consumption? Or perhaps better, why do we feel like the current practises are largely not sustainable? Many who speak of sustainable agriculture mean for crop cultivation to be free of pesticides and genetically modified organisms, where animals are free to roam to pastures as they please, and where the harvested products are not processed with industrial chemicals, such as preservatives and artificial colourants. The process of preserving food by pickling it in vinegar does usually not fall under this notion of 'chemicals' as it is often seen as authentic and natural, even though it usually occurs in similar industrial facilities. Pasture-roaming livestock might also seem idyllic and natural but there are often still many problems associated with it. Animals are often forced to stand in the blistering sun because their pasture contains no trees or other protection from the heat. These pastures often form a monoculture of ryegrass devoid of other plants or insects, and the term "pasture-raised" does usually not imply that calves stay with their mothers for more than a mere brief moment.

Would an ideal agriculture system constitute as sustainable? One where pastures are fitted with trees and flowers and other crops, even though it reduces the profits of the farmer? One where calves would not be separated from their mothers, although it reduces milk production. One where the cow, at the end of this idyllic scene, sees its life ended by a non-threatening machine that ends its brain activity faster than its neurons can signal for pain. Is this sustainable? Well, as far as the life of the cow and the rest of this utopian farm goes, my answer would be 'perhaps, yes'. But in this idyllic hypothetical example we have not yet touched upon the enormous sprawling effects of livestock farming and the gargantuan inefficiency that is raising beef compared to eating plants. The direct methane emissions from farm animals and the indirect emissions from land clearing for the production and transport of the animal feed mean that most farms, including many organic ones, have more livestock than they can feed from their own land.

Alright, no more meat, we go vegan. But although this removes many of the negative effects of animal husbandry we have still transformed the land from its original old-growth forests, redirected the rivers for irrigation and changed the occurrence of species in the area. The objects that inhabited the spatiotemporal domain before the dawn of human agriculture saw a massive shift in how they attained their wants and needs. Some objects possessed the capability to learn quickly enough, allowing them to fit this new reality they faced. Some did not. Deer and rabbits are still roaming wheat fields as if they are clearings in the forest, that is, if industrial equipment and pesticides do not bring about their demise. Wolves still hunt these deer and rabbits, if there is no highway or hunter nearby to end them. But many objects have perished, and not just species. Intact topsoil, for example, crafted through a multitude of physical and biological processes over the course of millennia has been wiped away through the action of tilling, often long before the introduction of mechanised equipment. But boars dig in the ground as well, effectively tilling the soil. And when boars are not influenced by humans, they provide a welcome ecosystem service of breaking up hard patches of soil where later many different plants and animals can thrive. Tilling by itself it thus not an inherently bad practise, it seems.

No matter what humans do with regard to agriculture, one can always find an analogy from another object – boars tilling the soil, beavers redirecting streams and rivers, lightning and drought burning forests, rain and wind flooding land. It is not a matter of 'what' human agriculture does, but rather a matter of 'why?'

Why? Because the impact is unavoidable. The only thing agriculture can do is choose which objects to impact, when to impact, and in what way – by changing farming to organic practises, by planting specific crops in specific seasons or by watering a field amidst a drought. And choosing implies a conscious decision. Hence it becomes a matter of why. So in our quest to find an answer as to what constitutes as sustainability with regard to agriculture, we have to investigate this choice.

Our current question of 'why impact one object in a different way to another?' is not synonymous with the aforementioned ethical question of 'how much good can counter the bad?' Naturally, attempting to ask ourselves why one would impact other objects, thoughts might drift to solutions how to minimise impact. This could very well be the beneficial outcome of people thinking about why they would impact certain objects. But it is not the fundamental answer we are searching for.

Let us also take a step back again from human agriculture, and focus on agriculture in the broadest definition. Here, I will use agriculture to mean the attempt at control of an object to provide its own wants and needs through providing for another objects wants and needs. Allow me to use an example to clarify this definition. A squirrel is known to collect nuts and bury them at various locations to hide them from his competitors, such that he can come back later throughout the winter months and have a food supply at hand. At first you might consider this to be hunting and gathering, as the squirrel goes in search of nuts instead of having planted an orchard. But is this correct? It is known that squirrels typically hide so many nuts that they often forget where some are, or they simply have hidden a surplus that they did not have to use. As a result, these surplus and forgotten nuts are in the ideal position to sprout into new trees, as having been buried slightly yields better roots development than nuts having simply fallen on the surface of the forest floor. And given that squirrels, like many species, have a territory they live in, I would argue that they engage in agriculture by cultivating more trees within the area that they choose to inhabit. The squirrel needs nuts as a source of food, but sacrifices some of this food to provide for the wants and needs of the trees, the forest, and other objects that link to them.

Once you start observing, I am certain you will find this behaviour in every object. And this also logically follows from our first four rules of sustainability. Every object is inherently linked to other objects, and although it has the freedom to pursue its own wants and needs and can learn how to do so and maintain an optimum, it needs to be attuned to the wants and needs of other objects and the rate of change of their ability to learn in order to avoid adverse effect to itself. Agriculture is thus nothing more than an object sacrificing some of its own wants and needs to other objects from which the first object expects a return at a later stage. It is the well-known adage of 'I am capable of giving you something you need now, such that you are capable to give me something I need later'. And from here we can derive the next rule of sustainability.

(6)

An object, of whatever form, utilises agriculture, understood as an object providing to the wants and needs of a receiving object such that the latter is enabled and expected to provide to the wants and needs of the former in return, as a means of providing to its own wants and needs.

The complexity of coupled objects

With our sixth rule of sustainability formulated, I want to explore the practicability of sustainable human agriculture in accordance with the rulebook developed so far. There is a multitude of labels and standards, all claiming to enforce sustainable agriculture, so I will highlight two that support the greatest level of sustainability. These are the Demeter label, and the 'do-nothing farming' approach.

The Demeter label was conceived nearly a century ago on the basis of the work of Rudolph Steiner. These formulate the basis of biodynamic agriculture which treats animals, plants and other objects such as soil and water as a single system. In practise it bans the use of artificial fertilisers, pesticides and industrial pharmaceuticals. It strives for a closed-loop nutrient cycle within a farm, meaning that the crops are fertilised by the manure from the livestock and the livestock feeds as much as possible on the crops. Often it also involves mystical beliefs and pseudoscience such as the use of astrology in the planning of sowing cycles and the 'embodiment' of cosmic energy into the soil through the burying of a cow horn filled with quartz. Although scientific insights vary on the efficacy of biodynamic agriculture compared to different forms of organic practises, I consider it the best quality label regardless as products that wish to display the Demeter label also need for instance a Skal organic certification. Demeter stipulates additional requirements on top of this more regular organic practise, such as the demand for roosters among laying hens and the ban on cow horn removal. The Demeter label is thus an enhancement of the sustainability of a product, rather than a claim of sustainability in its own right. A food displaying both the Skal and Demeter labels is thus better than a food that only utilises the Skal label, even if it would only be marginally.

Then there is the 'do-nothing farming' approach, which was developed largely by Masanobu Fukuoka in the last century. Although its teachings are not encapsulated in a quality label, there are farms across the world that claim to operate on the do-nothing principles. These principles are rooted in a belief that farming should follow nature as closely as possible and that human interference is not only unnecessary but often times also undesirable. The strongest example that Fukuoka gives in his writings is that of the rice field.[32] He developed a cropping

[32] '*The dragonfly will be the messiah*' and '*One straw revolution*' both extensively detail the farming practises developed by Masanobu Fukuoka, focussing predominantly on rice cultivation but also extend into other crops such as fruit and vegetables.

system where the residue from one crop is left on the field as if it were plants that naturally dried out at the end of their life cycle, which in turn blocks several weeds from sprouting and which creates a nutrient-rich local environment for the next crop. Further teachings from Fukuoka explain how to cultivate vegetables as if they were wild plants and how to operate a fruit orchard in as wild a manner as possible. Do-nothing agriculture is somewhat of a misnomer, as it stresses the point that it still requires significant work. It is by no means a lazy man's guide to farming but rather a philosophy where nature is left alone as much as possible to do as it sees fit. And although no formal guidelines exist for the do-nothing practise, pesticides, artificial fertilisers and pharmaceuticals are typically not used.

In some ways, the Demeter label can be seen as a formalisation of the do-nothing practise. Both approaches claim to work with nature rather than use it solely for human gain. Both avoid the use of man-made inputs as much as possible. Both aim to adapt to local conditions as much as possible. But both still miss a base rooted in OOO. Demeter cows can still be raised in an artificially created pasture and be forced indoors during the winter months. Are we sure that we gave all coupled objects, the cow, the barn, the field, the birds, enough consideration? Do-nothing agriculture can still cultivate a rice field where there was once a surplus of grasses, clovers, ferns, mice and owls.

A most pressing example is the Demeter certification of the cultivation of various species of potato, beets and cabbage in the Flevopolder in the Netherlands, of which the Oostvaarder-splassen discussed in chapter 2 are a part. The construction of this polder might very well have been classified as ecocide had the term been accepted by the Rome Statute.[33] The polder is the final culmination of the damming of the Zuiderzee into what is now a freshwater lake and a massive land addition for an already heavily poldered country. Little to no care went into accounting for all objects that depended on the saltwater inland sea and the area which is now a polder, let alone the object of the sea itself. Although for sure it is better to conduct organic rather than harmful agriculture practises in this polder, the mere fact of its existing and being utilised for human economic gain as extensively as it is, does not qualify as sustainable from my point of view.

I want to investigate these agricultural practises in more depth to see what is actually going on. Take for instance the phenomenon of community-supported agriculture, where a farm is supported through a combination of resources from the surrounding communities, be it through financing, offering expertise and insights or through physical support on the farm. These farms are typically praised for having their roots deeply embedded in society and human culture. The community has a say in which crops are cultivated and as such, the individual object of the farm instead becomes the collective object of society farming their sustenance. These farms also typically use organic practises, if only for the simple fact that people tend to disapprove of pesticides when they see for themselves what

---

[33] In June of 2021, the Independent Panel of Experts of the Stop Ecocide Foundation drafted the definition of the term 'Ecocide', explaining what constitutes as significant damage to the environment. Many years prior it was however already part of the Draft Code of Crimes Against the Peace and Security of Mankind, the document which would later become the Rome Statute. But several countries, among which the Netherlands, opposed this and ultimately caused it to be removed as a recognised crime.

effects they have on the land. So through the direct involvement of the consumers, the farm receives support to operate on organic-farming principles, which otherwise involve difficult pathways through legal and financing institutions.

This also applies to the Demeter quality label. Although products utilising the label are typically more expensive than their chemically and industrially produced counterparts, they still find their way to the market, largely through cooperative supermarkets– owned and operated by communities of consumers, by which they can avoid the market caps that are conventionally placed on products by for-profit supermarkets. By avoiding these market caps the cooperative supermarkets could have the industrial products for a lower price, but instead they opt to spend the surplus on the more expensive, yet more environmentally friendly products.

Both community-supported agriculture and community-owned businesses have a relatively strong base to operate in a way which is perceived to be sustainable. Although I am not keen to argue with the environmental benefits from these conducts, there are several downsides which cannot be ignored. For the Dutch context, ecological community supermarkets have a rather large number of members but these tens of thousands of members are still severely lower than the millions of clientele that visit the conventional supermarkets. Similarly, community-supported agriculture has grown explosively since the 2000s, but still barely chips away at the supply provided by destructive industrial agriculture.

Why is this? To my regret I must admit that for far too long I believed that this was a human error. That people were either too lazy to offer their own effort to a common goal, that they were lacking the financial or physical means to do so, or that they were simply not aware of these initiatives. And although these are certainly part of the explanation as to why sustainable agriculture has not yet gained the upper hand, they do not paint the entire picture. But based on the sustainability rulebook we can get a fuller picture. In order to do so we need to direct our focus to the word 'organisation'. We have said that each and any object interacts with its surroundings in accordance with its wants and needs while not causing too much damage to others. And that it can and wants to learn to find new ways to do so, while also being cautious of too much or too fast a change. Well, this all requires a series of objects that are linked together to form the coupled object of sustainable agriculture.

But industrial agriculture as a coupled object has become the dominating standard through its own series of linked objects. These are – albeit not exhaustively – market dynamics, public attitudes and established relations. Market dynamics require monocultures and pesticides through capitalist-driven profit maximisation and the established infrastructure. Public attitudes, through the lack of insight into the downsides of pesticides, favour economic gain over general well-being overall.

How did the objects at the base of industrial agriculture manage to link together and maintain their dynamic, while the objects required to form sustainable agriculture as of yet struggle to form a self-sustaining coupled object? Observing the coupled object, it seems that – perhaps counter-intuitively – the industrial agriculture coupled object is far more complex than its organic agriculture counterpart. Certainly, organic agriculture requires many additional certifications making that element more complex. But industrial agriculture has logistics chains of far greater complexity. This is primarily due to the fact that industrial agriculture requires more steps. Organic agriculture is focused on letting nature do most of the work whereas the industrial version requires inputs of materials and energy, processing of outputs, profits for shareholders. There are far more operations required within this particular coupled object, making it more complex overall.

And there is stability in complexity. The more complex a coupled object becomes, the more stable it can maintain its own self. We can use the human brain as an analogy. The brain is, in its essential form, a bundle of cells, primarily neurons. These are capable of some degree of movement in their search to connect to other neurons. It is these connections that form an important part of the functioning of the brain as a whole. But a mere two connected neurons naturally does not offer much of a basis for thought. It is rather the number of connections that creates a whole that is greater than the sum of its parts. After an individual has suffered a cerebral infarction, a brain that already had more connections before the infarction is better equipped to recover after the event. This is due to undamaged parts of the brain being able to develop new neurons and connections and in doing so take over functions of the damaged parts.[34] More neurons and connections thus leads to a better ability to maintain the coupled object of the human brain, even when faced with objects, such as a blood clot that caused the infarction, that try to undermine its existence.

The same holds true for industrial versus organic agriculture. Industrial agriculture has at present many more objects within

---

[34] The 2019 paper by Sharon Flora Kramer et al. titled '*The Impact of Physical Activity Before and After Stroke on Stroke Risk and Recovery: a Narrative Review*' describes how physical activity increases neuroplasticity (the ability of the brain to create new connections) which in turn can reduce the damage done by a stroke.

its coupled object that also see many more links between them. If one link loses its function, such as a commercial entity going bankrupt, other objects can grow quickly to take up this role. Even when significant damage occurs, such as an entire region being flooded, decimating crops and livestock, commercial entities such as insurance companies are relatively quick to provide the means to rebuild.

Do coupled objects strive for complexity, given that it is beneficial to meet its own wants and needs? The answer, I would argue, is 'yes' and the proof is easily found in another example: biodiversity. Biodiversity exists as a coupled object consisting of numerous individual and collective objects, those being kingdoms, species, herds and individuals. Complexity for the coupled object of biodiversity thus exists as large number of species and individuals, and the high degree of connectivity between them. Over the course of evolution, the amount of biodiversity has increased substantially. Additionally, life is no longer confined to the oceans. Many more species have developed, spread to all conceivable extremes of the planet and adapted to thrive there – from freezing cold to boiling hot, from barren deserts to floodplains, from anaerobic to aerobic conditions. In other words, biodiversity has become much more complex over time. It is certainly true that the extinction of one species can affect many others. But overall, the niche left by the extinction of one will quickly be occupied by others. The coupled object of biodiversity is, as I would argue, nowadays much better equipped to handle adversity than it was three or four billion years ago. This actually provides the basis for our next rule of sustainability. While developed within the context of agriculture, it can be generalised to become applicable to all objects.

(7)

A coupled object, existing as a collection of linked individual and collective objects, strives for complexity in order to create a stable space, safeguarding its wants and needs from external influence.

Circling back to industrial versus sustainable agriculture, what is required to for the sustainable version to gain the upper hand? We have now found that we should increase the complexity of sustainable agriculture if we want it to be able to compete with the current industrial standard. Naturally this includes copying certain elements from that practise, such as financial institutions and supply chains. But it would be of greater interest to revisit the do-nothing agriculture practise and Demeter label to see if they fit the complexity argument, and investigate how to make them more complex.

Recall that complexity is defined as the number of individual and collective objects and the number of links between them. And on that front, something interesting is actually happening. You see, for both agricultural practises of do-nothing and Demeter, most objects are identical to those of the industrial variant. There are consumers, farmers, retailers and wholesalers, processors, insurers and financiers. These can and should be developed further within the sustainable variant for sure. But eventually they are to match those of the current industrial object. From this perspective it would appear that the maximum achievable complexity of sustainable agriculture is equal to the current practise.

But there is more potential for complexity to be found in do-nothing farming and Demeter than there is in industrial agriculture, especially on the natural and social fronts. This is due to industrial agriculture's inherent trait of removing its own linked objects. Individual objects such as species that depend on the agricultural land are eradicated through pesticide usage. Collective objects such as the soil system are weakened, and effectively made smaller, through the use of artificial fertiliser which is tailored to the economic crop but lacks effectiveness, or even poses harm, to the collective object. And it actively seeks to reduce objects in its supply chain, mostly through exploitation of workers. Farmers are typically working in close contact with toxins without proper protection; other workers see long days in unsafe processing facilities. In doing so, these objects are effectively detached from clean water and air, their own health and relationships, and thus from the means to meet their own wants and needs.

The Demeter label and do-nothing farming on the other hand not only maintain the already present links, but actively seek to improve and increase them. They aim to bring species back to the farmland, even if that means having to sacrifice a part of the harvest to them. It means allowing crops to grow at their own pace, even if it may lead to less produce. It allows for the soil system to be as it sees fit, allowing for less nitrogen, phosphorous and potassium yet also more minerals for crops. It allows for cows to keep their horns, pigs their tails and poultry their beaks, such that they can live a life closer to how their biology sees fit. Demeter also uses additional guidelines, which are not part of the core certification and are as such not mandatory, that instruct farmers how to keep a calf with the mother cow.

These examples of Demeter and do-nothing underpin the second rule of sustainability, that an object should be allowed to pursue its own wants and needs. In doing so, they allow for complexity by increasing the links between livestock, crops, soil and more. Demeter-certified chickens that roam outside and keep their beak have the capacity to eat worms directly from the soil. By removing some worms, the soil system has an available niche that is then filled with different soil organisms, which in turn influence the dynamics above and below ground in their own right. And do-nothing fields that remain untilled keep the near-naturally present soil microbiome, 'near' because we should acknowledge that the microbiome is also influenced by above-ground activity.

Both Demeter and do-nothing can however increase their complexity even further by solely utilising local conditions. For instance by only using endemic or naturalised species. These species have developed over their evolutionary time frame to fill a very specific niche, with linked objects of their own. Endemic or naturalised species are thus part of the highest achievable complexity in a given region.

Another way to increase complexity is by relying solely on seed saving and by avoiding the use of F1 hybrids.[35] These are highly uniform, resulting from an industrial attempt at reducing links. When one individual object is copied and pasted, the overall farm then lacks diversity and links between individuals and their surroundings.

---

[35] The term *F1 hybrid* refers to the first offspring of two species that were selectively crossbred for one or more functional properties. For example, the savanna cat is a hybrid of the domestic cat (*Felis catus*) and the serval (*Leptailurus serval*). Hybrids are often sterile, meaning that they cannot reproduce. As a result, farmers who cultivate F1 crops are ultimately dependent on seed production companies to provide seeds for the new year. An increasing number of actors within the sustainable agriculture domain have been pushing back against this by providing natural seeds of which the resulting crop is capable of growing its own seeds which can be used by the farmer in the following year.

Basically, I am arguing for a scheme of 'give what you can; take what you need'. Every object will find its own way to meet its wants and needs. It becomes a matter of positioning yourself such that objects see it in their best interest to provide you with what you need, because you will give them what they want and need. A cow, for example, wants to take care of her newborn much the same way a human mother wants. Allowing the cow to do so then yields in return a calf that is less prone to disease and that shows a calmer behaviour.

But beyond livestock and generic crops, I want to give some special attention to staples. More specifically the global staples of grain, maize, legumes and tubers. These are the crops that are most often cultivated on large monoculture farms and are not widely available with Demeter labelling. Clearly these are more difficult to cultivate in a sustainable way, at least economically.

But why? Grasses, together with tuber growing plants, dominate the planet's landscapes, especially in steppes, savannas and tundra. These landscapes have formed naturally. The only thing that humans have done is select a handful of grasses that have been deemed edible. And since then these steppes have transformed from a mix of many grass and tuber species, to a mere few.

One can easily find a solution for this expansion of monoculture cultivation. Either more different plants need to be allowed to grow in between the wheat and maize, even if this results in lower yields because of the grassland having its own wants and needs. It could alternatively happen by increasing the functionality of these additional plants to humans, such as hemp for building insulation and flax for fibre, leading to a higher variety in species planted. But either way, a higher variety of species is needed because it would increase the complexity and hence create a more robust system.

Overall, there are numerous methods to increase the complexity of sustainable agriculture. I have merely provided a few examples. As long as human society focuses on incorporating as many objects into the agricultural system as possible and increasing the number of links between them, rather than decreasing them, sustainable agriculture will grow into a robust coupled object that can stand its ground against the pressures of the industrial agriculture coupled object.

4

Industrious species

Can we stop creating?

Humanity is a toolmaker, that is what I have always been told. As early as palaeolithic times, *Homo sapiens* has crafted wood and stone into items such as knives, clubs, baskets and spears. These were not only for hunting animals and collecting food, but also to fight each other.[36] Many hold that this toolmaking is what sets humans apart from other animals.

But there is much wrong in this view. Toolmaking and tool use are widespread among species. Several of them forge weapons from sticks and stones and in the realm of abstract

---

[36] Although the use of tools as means of violence in early human history is still heavily debated among scholars, some cave paintings (for instance one in Cova del Roure, Valencia) on the Iberian peninsula dating back around 10.000 years are thought to depict people attacking each other with bow and arrow. Regardless of these ancient times, the tools of humans are a common attribute in violent conflicts in present times.

tools create and operate rudimentary forms of an economy where one can buy favours with food or services.[37] As mentioned in chapter 2, crows use cars to open nuts. They drop the nuts on an intersection, let cars crack the nuts by driving over them, wait for the cars to stop at the red traffic light, and then eat the cracked nuts. Ants operate a type of healthcare system which provides amputation surgery. When an ant is injured, for instance after fighting, another ant will tend to the wound by biting the injured leg off. Although this might appear cruel, also because some ants still die after the procedure, it was proven that these amputations increase the injured ant's chances of survival significantly.[38] And on the subject of healthcare, there are several species of fish, such as the bluestreak cleaner wrasse (*Labroides dimidiatus*), that offer teeth-cleaning services to many different ocean dwelling species, even to species that could theoretically pose a risk to them. But these 'patients' do not seem to eat their 'dentists', as if they understand it would remove their ability to get their teeth cleaned. Clean teeth, with their subsequent ability to keep a healthy life, are apparently valued higher than a quick meal.

Creating does thus not need to be limited to physical, tangible items. Healthcare systems, types of governance and food distri-

---

37 '*Engineering skills in the manufacture of tools by wild chimpanzees*' (2025) by Alejandra Pascual-Garrido et al. describes various tools created and used by wild chimpanzees. And the paper titled '*Female Adelie Penguins Acquire Nest Material from Extrapair Males after Engaging in Extrapair Copulations*' (1998) by Fiona Hunter et al. describes observations that suggest some female Adélie penguins engage in a form of prostitution to obtain nesting materials.

38 As evident by the study '*Wound-dependent leg amputations to combat infections in an ant society*' (2024) by Erik Frank et al.

bution are all examples of how objects have created something that helps in providing their wants and needs.

I however framed the title of this section as a question for a reason. It appears to be interesting to pose the question if can stop creating since it seems that many objects do not continuously create whereas humans appear to continue creating for a multitude of reasons. It could be for a gain in efficiency, out of curiosity or boredom, or out of spite. Could it be something more? Many objects appear to create only until their wants and needs are satisfied, and then continue utilising the creation without significant modifications. They appear content with the current way of operating their creation. It is thus interesting to investigate the different reasons why objects might create, and if these reasons go beyond basic survival demands.

We can investigate this on the basis of emotions. Let us start with the assumption that emotions are universal across all objects. Based on the works by Paul Eckman or Robert Plutchik, several ways of defining emotions have been developed.[39] For our sake, let us use the basic emotions described by Robert Plutchik's wheel of emotions: ecstasy, admiration, terror, amazement, grief, loathing, rage and vigilance.

I will start with terror as a basis for creativity. All objects live with a certain degree of terror, typically with regards to other objects that have the potential to seriously harm their wants and needs. And as a result, they create as a response. A rabbit lives in terror of being eaten by a fox and as a result creates burrows for hiding from the fox. The fox itself also lives in terror of the

---

[39] Emotions defined by Paul Ekman can be found in chapter three '*Basic Emotions*' of the 1999 book '*Handbook of Cognition and Emotion*'. Those defined by Robert Plutchik can be found in the 1980 book '*Theories of Emotion*'.

weather. The fox also creates a den to hide from the wind and rain. Does the wind also experience terror? I would say yes. Wind is the result of air moving from a high-pressure zone to a low-pressure zone. As such it might experience terror of objects that block its movement. As a result, it creates an area of lower pressure around the blocking object which in turn increases its speed, based on the Bernoulli principle. So overall, one might argue that every object creates out of terror, with these creations taking many different forms.

Terror appears to be an emotion that spurs creation on the basis of basic survival needs. Recall, however, that we are investigating if objects create for other reasons, that go beyond matters of life or death. Reasons of grief, for example. It has become ever clearer through research that animals mourn their dead. Either through grievance rituals, such as elephants pausing at the location of a deceased kin.[40] Or through burying or cleaning their dead, such as is done by chimpanzees and dolphins.[41]

Species of the animal kingdom do indeed create on the basis of terror and of grief. But what about other, less obvious objects? Recall for instance our example of the mountain. If a mountain observes terror and grief, then it should also have a 'creating'

---

[40] A great review paper detailing the mourning behaviour of elephants is '*Elephant behavior toward the dead: A review and insights from field Observations*' (2019) by Shifra Goldenberg et al.

[41] '*Tool use for corpse cleaning in chimpanzees*' (2017) by Edwin van Leeuwen et al. describes the observed behaviour of a female chimpanzee cleaning the body of a deceased male. And '*Nurturant behaviour toward dead conspecifics in free-ranging mammals: new records for odontocetes and a general review*' (2016) by Melissa Reggente et al. summarises observations of several species of dolphins handling deceased juveniles.

response to it.

Let us first define what constitutes a response. Because it might be easy to state that grief results from the death of a loved one or from an unrealised dream, terror might come from a perceived threat to an object's physical integrity, and so on. But in its essential form, a response to an emotion is the manifestation of the wants and needs of an object. Grief is the coming to terms with the discontinued ability to love the deceased loved one, to see them move through life. Joy comes as a celebration of wants and needs that have been met. Admiration comes from observing another object achieve its wants and needs in a way that spurs motivation in oneself, allowing for the latter object to follow a renewal of its pursuits. I would thus argue that every emotion serves the purpose of facilitating an object in its pursuit of its wants and needs. And the 'creating' response could better equip an object to fulfil these wants and needs.

Here it should be admitted that the creating response could also potentially harm an object. Think of a person becoming an alcoholic after being fired from their job, or refusing to eat after the passing of a loved one. In these examples, grief can lead to a form of self-destructive behaviour. But even so, the emotion serves the wants and needs of the object nonetheless. The main aspect involving this self-destructive behaviour results from a hyperfocus on a very narrow set of wants and needs. The person that resorts to alcoholism after being fired might do so to escape the stress of their financial worries. And they succeed in this, they indeed forget their financial struggles. The fact that this behaviour then leads to malnutrition due to lack of proper food intake, or a reduced number of friendships due to people avoiding the alcoholic's behaviour are consequences that are irrelevant to the object in the present as this is not part of its current wants and needs. Only when the worry that caused the spiral to begin with is healed, do its wants and needs change again and does the damage to these new or renewed wants and needs become clear.

In short, an object experiences emotions to help its wants and needs, and creates on the basis of these emotions irrespective of if these wants and needs help the object in the long term. Having concluded this, we now move back to the example of the mountain facing grief. We stated in our earlier example that a mountain wants to stay intact and that stones eroding off of it are essential in the face of the coupled object of stone, mountain and magma, yet essentially constitute a loss to the mountain. The mountain might thus grieve this loss and create as a response. Now think of a hillside with a stack of loose stones that have detached from the mountain and have since stacked up in a pile from a broad base towards a narrow topside, leaning against the cliff that it originates from. Although I cannot fully understand the object of the mountain, I may still argue that this pile is created by the mountain, perhaps as a result of grief, to keep the stones close by for as long as possible. Perhaps it is partly out of terror or vigilance against losing more rocks, by protecting the rock underneath the pile from further erosion. Or perhaps it is out of amazement for the number of stones lost while still existing as a mountain, as it might be a rite of passage for mountains that grow older.

Creating is a common response to emotions which makes an object better equipped to achieve its wants and needs. Creating, in a sense, is a similar tool to agriculture and learning. But why do some objects appear to create more than others?

Recall for instance our discussion on learning and inherited behaviour. It is clear that some individuals within a species are innovators. They are the ones that are the first to create something, for instance a new way to dig a burrow. When this is then perceived to be beneficial to objects in its surroundings, they in turn start copying this new technique and the creation spreads in a society.

But one can assume that for every object that managed to create a new technique to build burrows which turned out to be beneficial to its society, at least one other object, and likely multiple, may have created its own new technique which turned out not to be beneficial, or perhaps even harmful. This failed technique is not adopted by its society. We see this, beyond human society, in many other objects. Sometimes a bird nest falls from a tree because it was not made sturdily enough. Sometimes a fox den collapses, a grebe mating dance might go unrecognised, a bumblebee colony can get extinguished over winter if the queen did not select a suitable spot for hibernation. In short, not every attempt at creating will yield a result that actually benefits the object's wants and needs.

The notion we used before, that creating makes an object better equipped in achieving its wants and needs, can thus be redefined: creating might make an object better equipped, while it might also harm an object. We are then left with the question: Is there such a thing as 'good' creating, which would imply that an object can avoid 'bad' creating, or is the act of creating inherently both good and bad? What I am asking is if objects that create something that in fact helps their wants and needs are somehow 'better' than their counterparts that created something that harmed them. Or is it the act of creating in itself, that by mere chance ends up benefiting or harming an object, what benefits an object?

Essentially these questions boil down to a comparison between weighting the benefits of creating to the potential drawbacks on the one side, and an iterative learning on the other. To elaborate on this, let us follow a thought experiment. We have already concluded that creating is an inherent trait of objects. But if the harmful risks of creating would outweigh the potential benefits, an object might choose not to create. In this scenario, an object would remain exactly at its current emotional state, as emotions dictate creating. It would be a perfect balance between progression and conservation – no will to move forward, no desire to move back. One should immediately observe the issues arising when other objects, to which our example object is linked, are in fact continuing to create. In doing so, these other objects continuously change the object space which would force our example object to follow suit or perish. It will get eaten if its predator object learns a new way of catching it. It will starve if its prey object changes its habits or location. Only if every object in the universe freezes, can an object stop creating. The second that one object changes something seemingly insignificant, it creates a ripple that reverberates through the total object space and cascades every object into a state of creating until another perfect equilibrium is found.

I contend that there cannot be such as a circumstance in which an object does not create. The universe is never still; everything is in constant motion no matter how small and no matter the possible vastness of the timescale. Since an object cannot exist in a state of 'non-creating', there cannot exist a discussion on the benefits versus the risks of creating. The act occurs regardless. This is especially the case if one were to observe the act of creating as its own object, we can understand why creating is as persistent as it is.

Since creating can be considered its own object, the situation where an object is creating should thus be understood as multiple objects interacting. This discussion has thus become related to our third, fourth and fifth rules of sustainability, related to learning, attuning this learning to one's surroundings and balancing this learning with the negative aspects that it might instil in one's surroundings. Since an object that creates essentially interacts with this process of creating, both objects confine each other and in doing so, impart learning in each other.

Recall for a moment that an object learns by forming an optimum of its own wants and needs, which is bound by the limitations set by other objects. Furthermore, no object should attempt to change these limitations which are imposed on it too abruptly, lest it will face adversity, and every object possesses the mechanisms of progression and conservation to invoke this. Let us also bring the coupled object of learning into our discussion. So, an object interacts with the object of learning, creating a coupled object. And all three of these objects, the two individual objects and the coupled object, possess the ability to learn, seeing as they are objects and thus follow the rules of sustainability. And this yields the highly interesting situation where the act of creating is bound by the confines imposed by the very object that does the creating.

What this implies is that the object that does the creating and the act of creating itself are both bound by the same rules of sustainability. If we would now thus establish a next rule of sustainability involving creating, for instance that every object experiences emotions and creates as a manifestation of those to achieve its wants and needs, this rule then also applies to the act of creating itself. And what this implies is that the act of creating can by itself also create.

We have thus ended up in a circular pattern, where every object is indubitably involved in creating as they are continuously influenced by objects to which they are linked, where every instance of this creating yields a link to an object of the act of creating, and where each of these acts of creating influence all objects to which it is linked. The act of creating thus creates a further act of creating.

Now we can finally return to the discussion of whether creating always benefits an object's wants and needs or if it could also harm them, and why some objects appear to create more – and 'faster' than others. We stated in our sustainability rulebook that an object is inherently linked to other objects to achieve its wants and needs, that it has the capacity to learn to optimise its own wants and needs, that is has the ability to perceive negative pressure resulting from when its own wants and needs impose too large a limitation on other objects, and that creating is an inherent trait of all objects. When combined, these rules essentially dictate that the amount, or the speed, of creating by an object depends on how many objects it is linked to. When an object is linked to many others, it can exert pressure on the wants and needs of a large number of objects, which then return a great deal of pressure that impacts the wants and needs of the primary object, and as a result this, the primary object keeps creating to maintain the ability to achieve its wants and needs. And thus we arrive at our next rule of sustainability.

(8)

An object, of whatever form, experiences emotions and engages, based on these emotions, in creating to achieve its wants and needs, where the amount and the speed of creating depends on the number of objects it is linked to.

The reactivity of coupled objects

Now that we have established our newest rule of sustainability and concluded the exploration of creating, I wish to focus to the practical implications on human society. Since there has been debate on the fundamentals of economic systems and the metrics by which to measure the progress of society, the sustainability of the industrious aspect of humanity is as relevant as ever. Such debates cover fundamental questions of capitalism versus communism, the free market and industrial output.[42] Societies deem industrial activity as key to stability. Unemployment sparks civil unrest; hence governments aim to facilitate the creation of jobs for as many people as possible.

Now, I am not going to participate in the debate of capitalism versus communism, as I have no preference for either system. I

[42] Since the establishment of the Millennium Development Goals (from 2000 until 2015), the subsequent Sustainable Development Goals (2015 until 2030) by the United Nations and the planetary boundaries in the scientific magazine of *Nature* (volume 461, issue no. 7263 on 24 September 2009), debates have been ongoing between proponents of 'green growth' and 'degrowth'. Green growth aims for a decoupling between climate impacts such as greenhouse gas emissions and economic growth, through means such as technology advancements. Degrowth on the other hands states that economic growth is the root of the environmental impacts of society and that is can hence not be part of a solution to these impacts.

will however not fully speak in favour of free markets and gross domestic products, as these are considered unsustainable by their very nature by many an expert in the field of sustainability.[43] Nor will I blindly argue for propositions such as circular or biobased economies, even though they are often viewed as more sustainable than current economic practices. Instead, I wish to break down the fundamentals of human economies to its essentials of creating, starting from the sustainability rulebook which we have developed so far, and see if we can develop behaviours that would make an economy sustainable from our current point of view.

We can start with the realisation that an economy is essentially the collection of what is created by all objects which within the boundaries of the economy. An economy is understood as more than the mere physical, tangible objects that are created and should indeed include non-tangible objects too. These include domestic work, organising birthday parties, a relaxing day on a bench in the park and ideological frameworks. It also includes objects such as rain, a sunny day or a singing bird as these objects directly influence creating by the human object. In short, an economy is the sum of all creating by human objects and those objects which are linked to human objects.

One basic question that I wish to solve, is how an economy can be considered unsustainable. Since more and more people in the present-day state that the way that the global economy operates is inherently unsustainable, I wish to understand what makes

---

43 The main ecology-related opponent to the conventional economic schools of thought is Kate Raworth with her model of the ‘Doughnut Economy’. It describes the level of wealth needed to provide people with sufficient quality of life while staying within planetary boundaries.

it so.[44] The easiest explanation has already been provided by the environmentalist movement, in that the global economy is unbalanced, favouring only humans and more specifically a small subset of them, while omitting the wants and needs of non-human actors and future human objects. This infers that an economy will be deemed sustainable if the wants and needs of all the linked objects are considered equally. And it implies that human creating is inherently selfish. We should investigate both of these statements, as I will argue that both miss the mark.

Let's start with the latter, which states that human creating is inherently selfish. It insinuates that creating by other objects is not from selfish motives. We have however already established that creating is a reaction to emotions. And emotions are inherently selfish, in the sense that emotions only cover the object that experiences those emotions.[45] Creating is thus a tool which is always aimed at benefiting the wants and needs of the object that creates. Human creating therefore cannot be deemed inherently unsustainable, as human objects simply abide the same rules as any other object, unless we would deem all creating any object as unsustainable, which goes against the notion that objects have created for as long as there are objects which we have discussed in the previous section.

Since creating by human objects cannot be deemed inherently

---

[44] 'More and more people' refers to an observed increase in prominence of the discussion between green growth and degrowth, which are explained in previous footnotes.

[45] The term 'selfish' should here be understood to mean 'solely and neutrally related to ones self', rather than the more commonly used term meaning 'a lack of compassion or empathy'. Someone caring for an elderly parent might do so out of compassion, or from feeling bad when seeing their parent not being taken care of. Both are two sides to the same coin.

unsustainable, we can start with our investigation on the equal consideration of the wants and needs of all objects within an economy. This statement is rather complex. Because it seems to insinuate that these wants and needs can actually be considered in the first place. But who could or should be doing the considering? If the answer is human objects, we again end up in the same argument as before: human objects, similar to any other object, are selfish in nature. Of course, with our second and fourth rules of sustainability we already concluded that any object is obligated to consider, to a certain extent, the wants and needs of other objects if it wants to meet its own wants and needs. But 'to a certain extent' means that an object does not strictly need to consider all the wants and needs of all other objects equally, as long as it considers them sufficiently. Likewise, an object can be metaphorically pushing against the wants and needs of other objects for as long as it can overcome the negative pressures that result from this pushing. Briefly summarised, an object can do as it sees fit to other objects, as long as it accepts the negative consequences.

This acceptance can be observed in the commodification of nature and the development of financial instruments and institutions involving them. By putting a price tag on non-human objects, the human object essentially states that these non-human objects can and should be impacted as long as the resulting benefits outweigh the costs. Something similar comes about from the science of environmental life cycle assessment

(LCA) or related methodologies.[46] By calculating for instance the greenhouse gas emissions of an industrial process, the essentials of LCA state that this emission is acceptable unless the costs of emitting, in the form of for example climate change or species extinction, outweighs the benefits. Even though it might seem sustainable to minimise the impact of economic processes on the environment, one would still impact it nonetheless.

The unsustainability of this scenario stems from a violation of the seventh rule of sustainability. The global economy, the sum of all creating by human objects, should first and foremost be seen as a coupled object. What we have already established is that any coupled object should strive to maximise its own complexity. But by minimising the impact of the economy, while still allowing for a certain number of objects to be impacted, one is maintaining a steady-state of a number of links and linked objects at best, and realising a significant decline of links and linked objects at worst. It does certainly not strive to maximise them.

The question that arises is how to match this seventh rule of sustainability, on the maximisation of complexity in coupled objects, with all the other rules of sustainability. Objects after all are required to consider the wants and needs of other objects for their own benefit and to avoid harm to oneself. But why would an object choose the route of benefit (the proverbial carrot) instead of the route of harm (or the stick)? The question here

---

[46] Life cycle assessment is a methodology used to determine the environmental impact of a product, process or service over its entire life; from the production of raw materials and manufacturing, to the use phase, and ultimately the waste disposal. The overall impact is often expressed in terms of costs, as there is considered to be a financial price associated with damage to the environment.

refers to the seemingly observable reality where an object only changes its behaviour when it experiences the negative pressure from the objects it is impacting. But is this really the case?

I believe that we must distinguish here between the choice of not harming another object because it helps oneself, and the choice of not harming another object because it is the 'right thing to do'. Usually, sustainability professionals refer to the term 'nature-inclusive'. Although the exact execution of nature-inclusivity has a lot of variations and depends on the person you are discussing with, it essentially refers to the attempt of including as much nature into your activity as possible. This could mean a residential development where the designer incorporates a lot of different trees in the streets and special bricks in the buildings where bats can reside, or a high-rise office building with a rooftop park and a green facade. The core of this notion is that sustainability is thought to come from colouring nature into the human creations, instead of letting human creations be coloured by nature. And one should include nature in human creations as much as possible, even if it requires more effort and resources, simply because it is the right thing to do. Nature has its own rights that we humans should respect after all, as it the prevailing sentiment among environmentalists.

These examples offer a valuable insight. Take, for instance, the special bricks for bats. These are installed into facades of buildings and offer protection to bats in a similar way that a crack in the walls of a stone cave would do. But human do not install these bricks with the intention of receiving a direct benefit from the bats. Usually these bricks are installed because it is believed that bats are an essential part of the local biodiversity and as such can be beneficial indirectly. Sometimes biodiversity and other species are even said to have a right of their own. And although these bricks are not hugely popular and widespread in the Netherlands, they are a creation of the human object regardless.

What I wish to highlight by this example is that there exist creations that primarily appear to benefit objects other than the object that does the creating. So is this a definition of sustainable industry? Doubtful, I would say. Humans have only started producing these bricks relatively recently, as a response to declining bat populations in the urban environment. The creation is following a negative pushback from the bat objects. Human objects appear to now create something that benefits the bat objects, only after first severely impacting them. One might then see the creating of the bricks as a way to relieve some part of the negative pressures that human objects pose on bat objects, sufficient to not destroy them so as to keep receiving their benefits, but not alleviating the pressures so much that it reduces the wants and needs of the human objects. The human objects in this example have, through the creating of the bricks, found a way to continue impacting bats so as to receive benefits from it.

Let us explore this notion of gaining benefits through impacting other objects because we can assume that sustainable creating can be understood as the gaining of benefits through positively impacting other objects, whilst avoiding negatively impacting them. And by combining this with the seventh rule of sustainability, it will yield a coupled object of an economy that maximises the benefits of all objects that are linked in the coupled object.

As good as this definition might appear at first glance, the principles of OOO quickly banish it to the realm of practical infeasibility. As one of the main principles of OOO states that an object partially withdraws from an observer which we have discussed in chapter one, the object that wishes to positively impact other objects through its act of creating can never achieve this desire as it can never fully comprehend what constitutes as positive to these other objects.

This is the tragedy of creating: Either an object will gain from negatively impacting other objects to the largest extent feasible, or it will fail to positively impact other objects due to a lack of understanding of those other objects. I believe that this 'tragedy of creating' is the basis for the human perception of the absence of sustainability in their economies. Whenever a person starts to realise that their actions are harmful to another object, they will attempt to shift their actions to actually benefit that other object. But in doing so, these new actions are either still harming the object, or they will instead negatively impact another object. It is a noble wish to only positively impact all the objects around oneself, but a foolish one nonetheless.

Yet however foolish any attempt to break the chains of the tragedy of creating might be, we must still attempt regardless. I believe that humanity's efforts at this attempt have inadvertently focussed on the wrong part of the tragedy. The environmentalists and other sustainability advocates have been restlessly attempted to indeed break the chains, whereas their attention should better have been directed at merely observing the chains. By doing so, they would have come to understand that these chains are nothing short of humanity's own making.

Try to imagine the concepts of positive and negative as two objects themselves. As any object partially withdraws itself from the understanding of others as per the fundamentals of OOO, the human attempt to determine positive and negative impact stalls on an incomplete point of view as to what they entail. And by subsequently attempting to act on these incomplete viewpoints from the false conviction of understanding, one can never fully resolve any issue, as the issue exists with the object that attempts to resolve it. Put in simpler terms, positive and negative impacting are myths. One can only simply impact. This also proves the first fundamental rule of sustainability, that any object requires interacting with other objects to achieve its wants and needs. Interacting in this sense cannot be designated as either good or bad.

By concluding this, it becomes clear that the coupled object of the human economy reacts very much in tune with the sustainability rulebook we have devised so far. The rate of change of the object's behaviour and learning shifts more rapidly than that of its surroundings. Simultaneously it is reducing its own complexity by removing linked objects. As a result, it experiences more and more negative pressure from its surroundings. It is precisely this negative pressure that causes more and more people within the economy to attempt to shift the behaviour of the coupled object. And it is likely that even more people will make similar attempts as negative pressure on the coupled object increases.

Although it now seems clear that the coupled object possesses a quality of reactivity at the hands of individual and collective objects within it, a more pressing question I wish to investigate is why some objects within the coupled object react differently than others. In the case of an economy experiencing negative pressures as a result of overly impacting linked objects, some people are quick to advocate for the wants and needs of these impacted objects whereas others have their priorities elsewhere. Some advocate for the environment, but the example works just as well with people helping the less economic fortunate, or vulnerable groups like children of the elderly. Although I am writing as an ecologist on the subject of nature and the environment, there is a long list of other fields of expertise that deal with their own urgent issues.

So why do some individual and collective objects within a coupled object respond differently than others to the pressures experienced by the coupled object? This has to do with the complexity of the coupled object as experienced by a constituent of it. We already established that a coupled object strives for maximum complexity – understood as a high number of linked objects that it consists of and a high number of links between these objects. But this does not mandate those links to be distributed equally across all objects. Some individual and collective objects can have a great many links to other objects, while others might not. Even if the number of links is equally distributed, not every object might be linked to the same objects.

If an object has a high number of links, it is more likely to be linked to an object that causes the negative pressures on the coupled object as a whole. It is therefore capable of directly observing those negative pressures. This is a fundamental aspect of direct versus indirect links within a coupled object. Take for example an object that is severely limited in its wants and needs as per the coupled object exceeding the surrounding capacity for learning, given the fourth rule of sustainability which stated that the rate of change of an object's behaviour should not far exceed the rate of change of its surroundings. As a result, this impacted object pushes back against the coupled object. It does so, not equally distributed across the coupled object, but to only a few individual and collective objects within the coupled object. These impacted objects pass on the direct impacts to other objects within the coupled object.

Think of a child jumping in a ball pit. When the child hits the first balls, these balls start displacing and pushing on the balls next to them. These ones start pushing on those next to them and so on. In this situation, the balls that the child lands on directly might be completely squished, whereas the balls at the sides of the pit barely move at all. The impact of the child jumping in the pit has indeed been distributed across the entire pit, yet unequally as the balls that experienced the impact first took the largest hit.

And here we have the core of our discussion. In the metaphor of the child in the ball pit, there exists complete linkage of objects. All objects in the coupled object are linked to all other objects in the coupled object. Even though some balls are impacted more severely than others, such a complete linkage means that the coupled object of the ball pit adapts nearly instantaneously to the impact emanating from the child jumping in. The less-directly impacted balls will displace to allow the more-directly impacted balls to move out of the way of the child, and they will fill the void left by any squished balls.

What this example shows, is that the speed of adapting to negative pressures is proportional to the number of links between objects within the coupled object. Negative pressures can come in all shapes and sizes, and appear to often only impact certain objects within a coupled object directly. These impacted objects can then pass on part of the impact to objects to which they are coupled, alleviating the burden on themselves. The objects experiencing the secondary impact then also pass on part of the impact to more objects, and eventually the impact is felt by the entire coupled object.

Conversely, when a coupled object is missing more and more

links between its constituent objects, any external pressure will be less and less likely to be partially passed on from the primary impacted object. This primary impacted object will thus feel a larger share of the negative pressure. And if this negative pressure is of sufficient magnitude to severely disrupt the impacted individual or collective object's ability to achieve its wants and needs, it can weaken a constituent of the coupled object and in doing so, weaken the coupled object itself.

We have now found that an individual or collective object that exists as a constituent of a coupled object, reacts to external pressures by attempting to pass part of this pressure to objects to which it is linked. We have also found that coupled objects are better equipped at achieving their wants and needs if they have more constituent objects, which are also further linked to others. If we combine these findings with the notion that any object creates on the basis of emotions, which are a result of external pressures, we can state that all constituent objects of a coupled object tend to create in different ways, given that they all experience external pressures in different ways. Finally, we need to combine this notion with the finding that an object does not create with good or bad intentions, simply from selfish desire to better achieve its own wants and needs. In doing so, we can create another rule of sustainability.

(9)

A coupled object, existing as a collection of linked individual and collective objects, responds to external pressures by passing them through its linked constituent objects, where the response time correlates to the number of links between constituent objects.

This rule encapsulates the interaction between the coupled

object and its constituent individual and collective objects, and objects external to it. A coupled object experiences negative pressure from external objects when its wants and needs encroach too much on their territory. If the coupled object however does not possess sufficient constituents and links between them, it will be slow to realise this negative pressure and learn to change its behaviour. During this time, the negative impacts emanating from encroaching wants and needs of the coupled object continue to harm the external object. Therefore, sustainability should be regarded as a coupled object that can quickly realise the harm it is causing to other objects, and accordingly learn to change its behaviour.

5

Reciprocalism

Science uses discretisation as one its main tools to study the world around us, which manifests in two ways. First, the many different scientific disciplines and specialisations; from exact (such as physics and mathematics) to abstract (such as sociology). And second, the need to determine research boundaries in order to confine which questions will be given an attempt at answering. Although this has the obvious advantages that studies become more manageable and results easier to communicate and implement, it has a very serious drawback. This is found in the lack of understanding of how each discretised scientific unit fits into the whole of reality. Humanity has become a collection of specialists, in which it is a rare occurrence to find a true interdisciplinary generalist. And although interdisciplinary researchers have gained more prominence academically, they are still confined to a subset of all scientific disciplines. Moreover, the vast majority of people are trained in one highly specific field of expertise. And although it seems that many claim to understand a vast array of topics and possess a multitude of skills, the reality is that this claim is more often based on false personal perception than on reality.

But this specialisation is actually neither new, nor an issue in itself. Although historically a larger portion of the population might have possessed basic skills like sewing, divisions of labour have always existed. Men might have been more often, although not exclusively, working the land or hunting while women maybe more often did tasks like sewing and cooking. There are also plenty examples to be found of women in positions that are traditionally attributed to men, and vice versa. This is not a take on gender roles but rather a demonstration of the ever-present specialisation of individuals within a society. And specialisation is most definitely not limited to humans. In bee colonies, some take up the role of protecting the hive, some scout the area, some collect pollen and others take care of the offspring. In many other species that live in groups, like wolves, guarding duties tends to befall on the largest and strongest individuals, whilst leading is usually left to the oldest and most experiences individual. One might even argue that any given complete species is a specialisation, as they occupy their own specific niche.

My point here is that specialisation of skills and knowledge cannot directly be deemed unsustainable, as it is widespread across objects beyond humans. We can also read the second rule of sustainability in this manner, where every object has the freedom to pursue their wants and needs in ways that they see fit. This leads to specialisation, as every object has, at least slightly different wants and needs and thus will acquire different skills and knowledge to achieve them.

What is a problem, however, is when an object – a human or human society in this case – observes another object and considers their observation to mean a complete understanding of the observed object. We have already established before that one of the fundamentals of OOO states that every object withdraws from other objects. Complete understanding of any object is therefore not achievable.

To contextualise this, we already investigated throughout the previous chapters that humanity's impact on the natural world is commonplace. Every object impacts other objects, sometimes too much, and then faces adverse pressures to rebalance the interactions. What is different in the case of human society, is that we are the objects that are impacting too much. The whole of humanity as a coupled object, through a portion of its constituent objects of engaged individuals and environmental groups, is thus aware of its own impact. And it is these environmental movements that work to change the impact from humanity before the adverse effects on our own coupled object that attempt to rebalance the interactions become too much. This is then seen as the coupled object learning to change its behaviour for a different interaction with its surrounding objects.

What these engaged individuals and environmental groups are doing, however, is observing at most a portion of the objects that are adversely impacted by humanity, claiming that they understand how they are impacted, claiming to understand how this impact should be mitigated and pushing the coupled object of humanity to change its behaviour accordingly. But as no object can ever fully understand another object, the changed behaviour from the coupled object or constituent objects therein might indeed result in a reduced impact on the observed objects while it could also increasing the impact on others. Or it might lead to a change in impact, but the impacted object still faces adverse pressure from humanity. And this is where the meaning of the term sustainability starts to deviate, even between the most specialised of human society. Some state that we should focus on chemical pollutants in water, or that we should reduce greenhouse gas emissions to the atmosphere. Others claim that we need to redistribute wealth as it otherwise leads to excessive consumption in a minor group of individuals.

This brings us back to the topic of scientific discretisation. Sustainability, at least partially, is the sum of interactions of all objects in accordance with the fundamental rules of sustainability, which we have established prior. But the approach of those who wish to stimulate sustainability, whatever definition they use for the term and no matter the good intentions, is to attempt to observe the wants and needs of every object – past, present and future –and account for them accordingly. This is practically infeasible. It will lead to certain objects being overly accounted for as they are observed more or are easier to understand, while others are undervalued, and the debate will continue as to which objects to include in the observation, and in what way.

Instead of building up sustainability as a sum of its parts, one should instead look to account for all interactions between objects in a continuous manner. This would account for all interactions equally, yet admit that no two interactions are the same; it would be a manner that allows for an unknown number of variables that has sustainability as the ultimate result. We can borrow a concept from computer science which can help us understand further.

This concept is the genetic algorithm, inspired by the process of natural selection and evolution. The basics of such an algorithm state that every participant operates according to a set of variables which are usually randomly generated to initially maintain all possible outcomes. These variables describe the fitness of an individual, in other words the proximity to the optimum outcome. All participants act and interact according to these sets of variables to attempt to find some optimum solution to whatever boundaries set by the researcher, which is typically not achieved within the first generation of participants. The participants then generate offspring with their own variables which are usually a combination of the parent's variables mixed with a small degree of random new variables, akin to a child learning behaviours from its parents in addition to generating a set of its own skills. The new generation of participants then go through the system again, attempting to find an optimum solution. This loop then continues either until a sufficiently stable solution has been found or until a certain number of generations has participated.

If we were to attempt to directly apply any genetic algorithm to our continuous function of sustainability, we would run aground before we would even start. This is the result of genetic algorithms requiring a set of variables that describe the fitness of an individual, which is the very thing that we are missing. If we would know exactly what each and every object's wants and needs are, we would be able to understand their most optimum way of interacting with other objects, and also with which objects to do so, in order to achieve the closest result to their wants and needs. But as we have already discussed, it is impossible for an object to understand the complete wants and needs of any other object.

In order to circumvent this issue, we should understand our genetic algorithm as a multi-objective optimisation problem, meaning a situation with more than one definition of fitness.

This is best understood as a Pareto efficiency problem. While the concept of Pareto efficiency is primarily developed for research in the field of economics, it can be applied to our case as well. The working principles of the Pareto efficiency describe a system in which a change is considered an improvement, and hence accepted, if and only if it helps at least one individual without harming any of the other individuals. And the ultimate goal is then to have implemented all possible solutions such that no further happiness can be created without harming anybody else – called a Pareto optimum.

By using a Pareto approach on our sustainability discussion, it would yield us an optimum point where the happiness of everybody, in the form of wants and needs met, is maximised and the suffering of everybody, in the form of wants and needs obstructed, is minimised. This sounds great, at least in theory. But in order to use the Pareto approach, we still run into the same issue as we have with the genetic algorithm, that is, the lack of possible understanding as to what constitutes as wants and needs met or hindered.

The mentioning of genetic algorithms and Pareto efficiency was however very much intended, no matter the practical limitations. Although no exact solution exists with which to determine the sustainability of something, it does not mean that sustainability itself does not exist. This also follows the withdrawal argument of OOO, which states that something can exist irrespective of my understanding of that something. Gravity has existed before the formal scientific denotation, gravitational waves have existed before their detections. My take is that sustainability exists even though we currently may lack the means to properly observe it. The aim of this book, or any other attempt at understanding the concept of sustainability, is thus not to influence the concept itself but rather our view of it. And this goes beyond sustainability; it holds true for essentially every aspect of life, even life itself. This is because when complete understanding is fundamentally impossible, learning becomes an infinite endeavour.

This observation links back to the rules of sustainability that we have developed before. They essentially state that objects exist at the hand of a constant dynamic, a push and pull, between the wants and needs of every object. It is a dynamic that is impossible to navigate perfectly, as no single object can fully understand this motion. But it is a dynamic that every object is required to follow as closely as possible, regardless.

That is the true fundamental definition of sustainability: *a dynamic and continuous interaction between the wants and needs of each and every object, an emergent property of a complex system, beyond the understanding of any single object.* It is something that dictates every movement of the system whilst simultaneously only existing because of the system. Sustainability is the mythological proverbial snake that eats itself, a contemporary circular economy. Without sustainability there could be no object space, without object space there could be no sustainability.

Now we can continue on the practicability of the subject. As I mentioned genetic algorithms and the concept of Pareto efficiency earlier, we can utilise them to bring sustainability into practise. I am curious to explore what it means for humanity to follow the dynamic of the object space.

To begin our analysis of the practicability of genetic algorithms and Pareto efficiency, we need to understand that they both depend on continuous learning. They work by making a choice, changing a variable, observing the outcome of the system and ultimately accepting or rejecting the choice based on the outcome relative to the wants and needs for a given object. Basically, they work like iterative analysis.

Genetic algorithms and Pareto efficiency typically exist as analyses performed by researchers or governments. I call this a centralised approach. They for instance want to understand if a set of choices they can make benefit the most amount of people, without causing too much harm. But this requires an object to be positioned outside of the system and again possess a degree of understanding of the wants and needs of the impacted objects, while in the case of studying sustainability our researcher is actually a mere part of the system. Typically analyses involving genetic algorithms and Pareto efficiency are performed based on a predetermined set of parameters which are considered to describe the wants and needs of the impacted objects. But in our case these parameters are not guaranteed to actually describe the complete wants and needs of objects. Quite the contrary, it is guaranteed that they do not, again following the fundamentals of OOO.

I propose a decentralised approach instead. This acknowledges the inability to completely understand the system and the objects within it, and instead allows every object to define its own parameters that describe its own wants and needs. It will however also shift the burden of responsibility from a singular object to all, as it requires every object to only accept those choices that benefit themselves without harming others. But this responsibility can already be found in the fourth rule of sustainability on the rate of change in behaviour which should be attuned to an object's surroundings which means that objects are capable of sharing in this decentralised responsibility. Such a decentralised approach where every object shares the responsibility to account for others is essentially an enhanced and extended form of reciprocity.

Reciprocity can be defined as the social norm, typically used in the context of sociology and hence applied to humans, that describes an equivalent action of someone in response to an action by someone else. Reciprocity can take one of three forms: direct, multilateral and generalised. Direct reciprocity is a one-to-one exchange. One person or organisation gives something to someone with the intent of being given back something of equal value by that same person, usually also within a known time frame. For instance, a person buys a book from a shop and pays fifteen Euro for it. The book is being given by the shop to the person, and in exchange, the shop directly receives fifteen Euro worth of currency from the person.

Aside from direct reciprocity, there also exists generalised reciprocity although I will later argue why I disagree with the existence of this form of reciprocity. Generalised reciprocity is considered an exchange where someone provides something to someone, without directly expecting anything in return. An example is often given as a person who donates blood to a hospital. This person receives a functioning healthcare system and perhaps a nice feeling after doing a 'good deed', but no direct repayment by the hospital is made to the donor.

Finally, there is a form of reciprocity that exists between the direct and the generalised types. This is called multilateral reciprocity, and works on the basis of a rotating system of donors and receivers. For example, one person receives a birthday present from a group of friends, and this one person participates in giving a present as a group at the birthday of another friend. It is possible for new people to enter this type of reciprocity – new friends that did not equally participate in the past exchanges, or people leave the system. It is always clear who gives to what and at more or less which value, as it is a present from the group to someone within it, for which usually a price is agreed upon. But there does not need to be a direct, or even fully equal, exchange between the participants. In that light the previous example of the blood donor should be seen as a form of multilateral, instead of generalised, reciprocity. This person has a clear intention for their blood donation, which is to have a functional hospital that can save lives, but the blood donation does not have to be of equal value to the function that the hospital brings to the donor, or to other patients for that matter. In this regard I will argue that generalised reciprocity does not exist and that it should instead be understood as a type of multilateral reciprocity. One always gives in order to receive something in return, even if that something is of less value than what was given. And even if the return comes after the death of the providing object, for instance in the case of a person building a cathedral because he wishes his great-grandchildren have a place of worship.

Now that we have these forms of reciprocity established, we can proceed. The contemporary human society operates on something in between direct and multilateral reciprocity. Some reciprocal relations are of the multilateral kind, such as birthdays within friend groups or parents volunteering at the sports club of their children. But most human interactions work on a basis of direct reciprocity. I work for a manager, providing a predetermined set of services for a predetermined amount of money as per my contract. With this money I buy products and services, providing a predetermined amount of money for a predetermined amount of a specific good or service. I do send a small portion of my received money to charities which I would classify as a form of multilateral rather than generalised reciprocity. I donate money with the intention of receiving a very specific benefit in the form of protecting the environment, researching medical treatments or providing poverty alleviation. I am not throwing money into the wind; I wish to receive a somewhat specified return for my provided services. Even taxes can be considered as a multilateral type of reciprocity than a type of general reciprocity. I pay taxes with the intention of living in a country that takes into account healthcare, defence, education and all other societal expenditures, preferably in a ratio that I agree with – more to education and less to defence for example. So even taxes have a multilateral element to them, as the benefits I receive do not have to be equal to the amount of taxes I have paid. But they do need to be of enough value to me, otherwise I will vote for a different representative next time, or participate in a protest. The amount of value returned in a reciprocal exchange deemed 'sufficient' depends on many factors, such as the value provided, the traits of the providing object, and so on.

There is always some requirement put on the exchange. And in such an exchange, any object will always provide resources with the intent of receiving a return. This return can then either be of less value, or of more value than that which the object has provided. A return of equal value is unlikely, as either object participating in the exchange is incapable of fully understanding what constitutes the value of the exchange to the other object, as per the foundational argument of withdrawal of OOO. And even if a completely equal exchange occurs, neither participating object will know of it. Whether an exchange yields a return of less or more, or even equal, value than what was provided by an object is irrelevant, as the important notion here is that any object will always expect some form of return.

This, instead of the direct and multilateral types, should be the single definition of reciprocity: an object always gives so that it can take. And conversely, an object only receives when it gives. For example, the precursor to the modern concept of economy, which was one of giving gifts rather than one of exchanges of goods, services or currency.[47] This gift-giving economy could for example be as follows: A hunter comes back from the hunt, in the middle of winter, with a massive stack of meat. It is far more than what he needs for himself and his family. He thus provides the stack of meat to the entire village community, where everyone can take as they see fit. As a return, the hunter receives berries from the gatherers in spring, and fish from the fisherman in summer. Perhaps somewhere during the year a labourer repairs his hut. All these exchanges are provided free of direct charge, based on the anticipation of receiving back when new supplies arrive.

This gifting system functions if a person in the community gives to the whole community; in return he will receive benefits from many in the community. But in this way, it does not account for reciprocity towards the mammoths, berries, fish and all other objects that participate in the exchange. Because if the fisherman catches a vast catch of fish to supply to the community, this community could grow, and as a result the fisherman could receive more goods at a later stage. But

[47] '*Sharing, gift-giving, and optimal resource use in hunter-gatherer society*' (2021) by Matthew Baker et al. describes the functionality of a gift-giving economy very well. Furthermore, '*Why hunt? Why gather? Why share? Hadza assessments of foraging and food-sharing motive*' (2022) by Duncan Stibbard-Hawkes et al. interviewed a contemporary hunter-gatherer group (Hadza people) and found their exchanges predominantly contain gift-giving and reciprocity.

simultaneously in doing so, the human community overextends their wants and needs into the domain of the fish community, damaging the fish population. This fish community can then fight back against the pressures of the human community by moving farther out into the sea or into deeper waters, hindering the humans' ability to catch them. The retaliation by the fish will however unlock a chain of responses by other connected objects, responding to a shift in wants and needs. Some deep-water fish will start hunting the new shawl that moved into their domain, birds that previously hunted the fish are now forced onto land to eat the harvest from the humans, other fish might take over the space left by the original shawl. The birds that will now eat the harvest intended for human consumption can be understood as a direct negative pressure on the human community as a result of the original pressure by the human community, even though it does not directly come from the impacted fish community. Had the humans prevented the fish from migrating to deeper and farther waters, these birds would still have their original source of food.

In this example, the human community did not give something before it took. Had it given something to the fish community, for example food or protection, the fish community could have grown to a level that would be able to provide food to more objects before it would feel compelled to retreat to safer waters. And had it given to the bird community, these birds would be less compelled to enter the human domain to eat their harvest. The logic here is that an object will always take something from you if you take something from some object, whether that is the same or a different object than that you took from. The only choice is in what to give for the exchange. If what is given for the exchange is of insufficient value, someone will take something that was not originally provided to the exchange.

I propose to see the matter of sustainability from the point of view of an enhanced and extended form of reciprocity. The enhanced and extended aspects of this refer to the understanding and intention behind an exchange. You can choose to take without offering something of sufficient value, but as a result someone will take something from you which you might not have wanted to offer to the exchange. Let us define this as an object's choice between a deliberate offer and a random offer. 'Choice' here might be a slight misnomer because an object might make the conscious decision to make a deliberate offer, which later turns out to be of insufficient value. The choice then refers to a supplemental offer the object should give. The providing object will thus be providing offers, either deliberate or random, until they have reached a combined sufficient value.

What about fairness though? The receiving object might actually receive something of sufficient value, while demanding more. Conversely, a providing object might provide something of insufficient value, but manage to avoid the resulting negative pressure and instead divert it onto another object.

Let us start with the first case, that of an overly demanding receiving object. If the receiving object has received something of sufficient value for that which it provided, the exchange should end. If the receiving object however chooses to continue the exchange and demand to receive more value, this should be seen as the inverse of the original exchange. In this situation, the receiving object is attempting to take and must provide something to the object that was originally wishing to take. An object trying to overextend the already sufficient value which it received in an exchange is then susceptible to the same issues we have already discussed. Overextension on the receiving end thus does not appear to be a complicated matter in its own right.

It does however become complicated when investigating the diverting of negative pressures arising from an underextending providing object. Recall the example of the humans, fish and birds. The humans taking the fish, without returning something of sufficient value, which was leading to birds eating the harvest of the humans. Take now the situation where the birds are eating only from one farm, ruining this particular farmer's livelihood. The objects that caused this situation, in this case the fishermen who did not sufficiently give for what they took, are then capable of buying the harvest from another farm. The fishermen thus might not experience much of a negative pressure, only that they have one less farm to choose from when buying produce, while that one farmer is utterly ruined. Even if one would factor in an increase of prices due to an overall reduced supply of farm produce, the additional costs are distributed among all who buy from the farm. Again, our fishermen do not face the bulk of the negative pressure resulting from an imbalance they themselves caused. In this situation, an equal give and take system might not appear attainable as one could take more than what was provided for and simply divert the negative consequences onto other objects.

To resolve this issue of fairness, many solutions have been developed throughout human history. The one most common appears to be that of societal norms. Although varying between communities, societal norms share of commonality of being instilled in young members of society and enforced throughout adulthood. Enforcement typically takes the form of soft power, such as socially excluding those who do not comply with the norms of the community and rewarding those who do comply. Although there are likely an unfathomable number of societal norms, from what I have observed there always appears to be an aspect of fairness involved. If you would help out your neighbours a lot, these neighbours are then socially obliged to return some favour to you. And if one member of the community tries to enrich themselves by scamming others, this community member receives harsh social punishment. As successful as societal norms are to attain fairness within a community, they can be unsuccessful between communities. The social norms of one community might clash with the norms of another. This can be on the subject of food, clothes, behaviour, life goals or any other imaginable aspect. Social norms often lead to misunderstandings between communities which can escalate into competition, tensions or worse.

Beyond societal control at fairness, modern attempts often feature institutional control. The social credit system in China is one infamous example. Although not as encompassing as often portrayed in Western nations, as it currently does not include personal scores for individuals, it is in place as a scoring system for businesses and government entities. It attempts to allow individuals and entities such as businesses to see whether or not a business is trustworthy. The idea is that you would not consider conning your fellow community member, because it would yield you a lower social credit score meaning other individuals and businesses would be less likely to interact with you in the future. Although it might have its merits in theory, the practise is prone to autocratic abuse. This is also the main issue with any such institutional control mechanisms. They allow a singular object to exert control over all exchanges. And as we have discussed in detail in previous chapters, an object that dominates its own wants and needs over another object is grounds for a negative return pressure.

And China is not the only nation with its attempt at fair exchanges. The USA effectively operates a somewhat similar system – the 'personal credit score' – which functions to provide information about a person who wishes to borrow from a loan provider who can judge the risk of not being paid back by that person. Although this is a system operated by commercial entities rather than government and hence evading the risks from autocratic abuse, it still functions as a one-way system. The borrower provides information to the lender, not the other way around. As such, the lender is the one who dictates to the borrower, effectively functioning as an object that overshadows another object with its own wants and needs – an unfavourable outcome.

A final example of fair exchanges is that of third-party verified quality labels. Labels such as the European mark of product quality CE, organic food labels Skal and Demeter, sustainable beauty care label ECOCERT Cosmos Organic and many more.[48] These are typically operated by a foundation or government entity, which in deliberation with stakeholders, comes to a set of guidelines that describe the quality, safety, sustainability or other traits of a product or service. Independent third-party organisations are subsequently contracted as notified bodies to verify the continued following of these guidelines by those who fabricate and market the products or provide the services. The aim of such third-party verified labels is to ensure that the buyer is informed about the ways of operating by the supplier and the quality of their product. It is a means to ensure a fair exchange

[48] Although CE is not officially an acronym, the European Commission used *Conformité Européenne* (French for European Conformity) in the press release '*European Commission welcomes European Parliament's approval of new rules for pressure equipment*' in 2014.

towards the buyer, rather than a fair exchange towards the seller as is the case with the American personal credit score as well as social norms in general. Both the American personal credit score and social norms mainly enable a supplier to judge buyers. The power to dictate the exchange thus favours the supplier in the case of credit scores and social norms, and the buyer more through quality labels.

In either case, I believe that true fair exchange is non-existent. Either the buyer or the supplier dictates the exchange; there never seems to be an equal judgement of either side by the other. An equal judgement is infeasible. Take the concept of an exchange, being a transfer of something tangible or intangible between objects. As it only exists by the grace of objects engaging in an exchange, the exchange itself should be seen as a coupled object. And as we have already established through sustainability rule seven, any coupled object strives for an increase in complexity. The coupled object of an exchange does so by including as many other objects into itself as it can. As a result, an exchange never consists of only two objects that transfer something between them as the transferred something is another object again, naturally. But by transferring that something, the exchanging objects influence other objects. Take for instance the hypothetical of me giving you an apple. Me, you and the apple are all part of the exchange. But by giving you the apple, I cannot give it to somebody else, or to something else, such as birds, worms or soil microbes. And as I have given the apple to you, instead of selling it to you, I now do not have money to send to a charity. All are thus part of the exchange, as they are all influenced by me giving you the apple. We can never hope to fully judge such an increasingly complex exchange on its fairness.

A tunnel focus on a fair exchange appears to be meaningless. But let us now recall the notion that an exchange should be seen as a coupled object. In addition to its drive for increased complexity as per rule seven of the sustainability rule book, a coupled object also responds to pressures based on rule nine which states that the speed with which a coupled object changes its behaviour as a result of external pressure is correlated to its own complexity. When combined, these two rules show that a coupled object becomes better at responding to external pressures over time when its complexity increases. If we aim for fair exchange, we should realise that a highly complex coupled object will be much better at withstanding pressures, such as arising from an underextending providing object, than a simple coupled object. A completely coupled object – possible only in theory – exists at a maximum where every individual and collective object is coupled to every other object. It will then even be able to instantaneously respond to such a negative pressure, directly steering it back to a fair exchange.

While a completely coupled object is unachievable in reality, any strive for complexity will enable a coupled object a faster response time to negative pressures. And when an exchange is considered to be a coupled object, the more complex the exchange, the fairer. To contextualise this, think for a moment about an unfair exchange. As an example, we use a thief stealing a bike at a train station, a well-known occurrence. The thief wants the bike and hence engages into an exchange with its owner. The value provided by the thief is zero, and is insufficient for the value of the bike which he takes. This is obviously an unfair exchange; the thief ends up with more value than the owner. But the owner might have insurance, which covers the value of the stolen bike, allowing the owner to purchase a new bike. Likewise, the police might catch the thief and give him a fine. Although everyone involved now seems to again have the value that they should have, we have still not achieved a completely fair exchange. This is due to the thief now having taken an amount of time and effort from police, insurance company and bike owner, which he has not yet given anything in return. For sure, this situation is fairer than a case where the thief is not caught. But a totally fair exchange would be one where the thief is deterred from their plan to steal the bike altogether. This could be achieved by for instance involving, aside from the additional objects of the police and insurance company from before, other visitors to the station who notify the police when they see the theft taking place. Or municipal counselling for the person compelled to commit the crime, to help resolve debts or addictions which might be the cause of such crime. The point is that by increasing the complexity of the exchange, its fairness increases.

To reiterate: An exchange, being a coupled object itself, strives to increase its complexity – as per rule seven. In doing so, it increases its fairness by limiting the time between an object providing insufficient value for what it takes and that object feeling the negative pressure from this action. This negative pressure will then, when of sufficient force, change the behaviour of the object that caused it. Therefore, for either party to experience a fair exchange, both sides should strive for as much complexity as possible.

This should be understood as a great many further exchanges, linked to the main exchange, rather than simply many other objects linked to it. This comes from our earlier realisation that an exchange by nature already includes many other objects besides the two objects that engage in the main exchange.

And depending where you cut off the frame of observation, any exchange can essentially include each and every object in existence. If you think, for instance, of a fisherman selling fish to a customer, the fisherman and customer are of course the main objects in the exchange. But this exchange also includes the fish, the boat and fuel used to catch it, the climate impacted by the use of the fuel, other marine creatures that can no longer eat the caught fish, plants that respond to an elevated temperature due to the impacted climate and so on. By incorporating other objects to an exchange, you automatically engage in new exchanges with these objects.

Each and every one of these new exchanges, naturally, in turn also strives for complexity. Essentially, once you start an exchange you will engage in an ever-increasing series of exchanges. This is how our definition of sustainability from before can also be understood, where sustainability is a continuous series of exchanges that strive for complexity in order to become fairer. And this ties back to our earlier argument on reciprocity, where you can only take something if you also give back. If I wish to exchange something with you, I first should provide value to another exchange in order to obtain that something. And even if I cannot judge completely what constitutes its value, the mere act of providing to one exchange already makes any connected exchange fairer. And thus, we enter into a loop where each exchange is made fair by the next exchange.

To visualise this, imagine a graph where the y-axis describes fairness, understood to mean the sum of the wants and needs of the objects that are part of an exchange; the x-axis describes the complexity of the exchange. The image below shows a possible curve for any given exchange. The function that we then draw starts at zero, the origin, being a completely unfair exchange, and moves to a horizontal line denoting the maximum achievable fairness by increasing the complexity. The speed with which this graph moves towards the maximum fairness will however diminish with added complexity, meaning the first added complexity has more effect on the fairness of the exchange than any subsequent complexity. And, due the withdrawal argument from OOO, our curve will move ever closer and closer towards the maximum fairness, yet it will never truly reach.

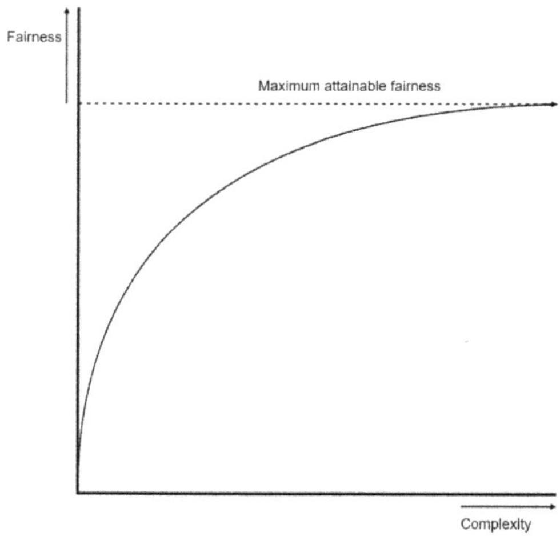

This we can now link back to our discussion on the subject of enhanced and extended reciprocity. Enhanced and extended refer to the connection between the main observed exchange, and the many subsequent connected exchanges, respectively. The attempt to include as many exchanges to the main exchange intentionally, rather than reactively, is what I wish to call reciprocalism. In other words, *reciprocalism is the act of achieving one's self-interest by aiding the self-interest of as many others as possible.*

Reciprocalism differs from conventional reciprocity in its direct benefits to both sides of an exchange. There is no waiting time between providing and receiving, no implied debt after receiving fish in winter for which you have to provide bread in summer. And there is no need to only exchange with objects which you deem trustworthy, or to hedge oneself through dedicated third-party objects against potential risks caused by unfair conduct by the other object of an exchange. The spread of risk and the insurance against unfair conduct are in the hands of both sides of an exchange. And through caring for themselves, exchanging objects directly care for those around them.

6

Complexity demanding

Reciprocalism allows you to increase the fairness of the exchange you are currently partaking in, by connecting new exchanges to it. It ensures that you get what you deserve based on the value that you provide, even if the object with which you are exchanging wishes to take more from you or provide less to you based on your provided value. This foundational principle of reciprocalism of aiming for fairness, does however only work for exchanges where either party has a position to treat the other unfairly.

To illustrate this, imagine that you are holding me at gunpoint and demanding my money. You and I are having an exchange in which I am giving you my money and in return you are not harming me. If I am to decline your offer on the basis of it being unfair, you will of course harm me. And so I am left without a choice, other than to accept your exchange offer. We examined a similar situation with the bike thief before. And likewise, in the current example I can still make the exchange fairer by ensuring that there is a decent and well-functioning police force to come to my rescue, or that there are neighbours who keep an eye on their streets and warn the police when they see that I am locked in a dangerously unfair exchange.

But this type of exchange, where you are holding me at gunpoint is not all that different from, for instance, a large corporation or national government exploiting nature or people in vulnerable socioeconomic situations. These people often have a mere choice between accepting the meagre offerings from the large corporation, or be driven off their land completely and be left with absolutely nothing. And this holds especially true for non-human objects, who often have little to no say in whether or not to accept the value offered by a human object in an exchange.

Although an exchange is made fairer by increasing its complexity, for both human and non-human objects who are treated in such a way by an overpowering object, it is wise to understand that the poorly treated objects often do not often have the means to link other exchanges to the primary observed exchange in an attempt to make that primary exchange fairer. A sub-Saharan villager being driven off his land by an armed group supported by wealthy nations might have little more to offer to other exchanges other than a plea for humanity. Westerners who might see a broadcast about the atrocities that befall this villager could feel sufficiently bad about their own inaction that they are compelled to send money to a charity to buy off their guilt. But our villager will only see any increased fairness to his primary exchange if his plea for humanity was sufficiently picked up by the global media and if it made people feel sufficiently badly about his situation that they would send money to a charity, and that this charity was in a position to actually improve his situation.

Here we see that fairness is at risk of becoming a commodity relevant to those who can afford it and do not need it, rather than to those who cannot afford it but who would definitely benefit from it. This is not simply a reiteration of the rhetoric of looking after those in need out of some moral stance. There is actually an aspect to it which benefits the self-interest, regardless of your own means. Earlier I claimed that each and any exchange basically links every object in existence. We can reconnect to the seventh rule of sustainability on a coupled object's drive for increased complexity. Logically speaking, it of course means that you are connected to the exchange which appears to only involve the villager and the armed militia. These things often appear to be sufficiently far away to not affect our lives, whether it is a war or conflict on the other side of the world, a destruction of a forest in a faraway land or any other unfair exchange that might not directly affect you. Each of these exchanges can and does affect you.

This is explained by the foundational principles of OOO, most notably, that an object is incapable of fully comprehending the reality of any other object – the withdrawal argument. To me, this can be extended into a realisation that the chain of effects that started with the unfair exchange in a faraway land, impacts objects directly connected to it, which in turn affect the indirectly connected objects. This process continues until the effects reach you. They might not appear related to the primary exchange, since they will have likely taken a very different form over time, but effects most definitely ripple through the spatiotemporal domain.

Naturally, what we are now considering the primary exchange between the villager and the militia, of course only came about because this militia exchanged with a delegation from the rich nation and perhaps other objects. Each actor gives the effects along to a new exchange. One could say that reality is in fact one continuously unfolding exchange.

Returning to the seventh rule of sustainability, we can understand this to mean that any human object which is in a position of power over another object should still strive for fairness. After all, the more links there are within any given coupled object, the more stable that coupled object becomes to negative impacts. Perhaps today this impact takes the form of militias harassing villagers in a faraway land. But tomorrow it may take the form of a breached dyke closer to home.

Stable coupled objects benefit all individual and collective objects within them. This I consider to be the frustrating reality of needing others. Sometimes it might feel better to disconnect from objects that appear beyond our sphere of influence. But it is not possible; either directly or indirectly, you are connected to other objects' exchanges. You subsequently directly or indirectly need these to provide for your own wants and needs.

An object does not exist in order to help others. An object exists because it helps others.

Fundamental rules of sustainability

1. An object, of whatever form, is inherently linked to other objects, and depends on interacting with other objects to provide its own wants and needs.
2. An object, of whatever form, should be allowed the freedom to pursue its wants and needs in ways it sees fit, whilst fulfilling its obligations to objects to which it is coupled.
3. An object, of whatever form, is capable of learning, meaning the forming of an optimum of its own wants and needs, adapted to the confines shaped by the wants and needs of objects to which it is coupled.
4. Learning, that is the sum of an object's ability to react to its surroundings and the behaviour it inherits from the objects it descents from, should maintain a rate of change to an object's behaviour which is attuned to that of the objects to which it is coupled.
5. A combination of progression (learning and adapting to changes in an object's surroundings) and conservation (the avoidance of the negative feedback from those surroundings if an object changes too fast), is used by every object as a means to achieve its wants and needs.
6. An object, of whatever form, utilises agriculture, understood as an object providing to the wants and needs of a receiving object such that the latter is enabled and expected

to provide to the wants and needs of the former in return, as a means of providing to its own wants and needs.

7. A coupled object, existing as a collection of linked individual and collective objects, strives for complexity in order to create a stable space, safeguarding its wants and needs from external influence.
8. An object, of whatever form, experiences emotions and engages, based on these emotions, in creating to achieve its wants and needs, where the amount and the speed of creating depends on the number of objects it is linked to.
9. A coupled object, existing as a collection of linked individual and collective objects, responds to external pressures by passing them through its linked constituent objects, where the response time correlates to the number of links between constituent objects.

www.ingramcontent.com/pod-product-compliance
Ingram Content Group UK Ltd.
Pitfield, Milton Keynes, MK11 3LW, UK
UKHW020152290926
15471UKWH00005B/825

9 798232 215798